ISSN 0044-2267

ZEITSCHRIFT FÜR ANGEWANDTE MATHEMATIK UND MECHANIK

INGENIEURWISSENSCHAFTLICHE FORSCHUNGSARBEITEN

BAND 60 1980 HEFT 11

AUS DEM INHALT

AKADEMIE-VERLAG · BERLIN

ZAMM Bd. 60 Nr. 11 S. 549—644 Berlin, November 1980

EVP 18,— M 34115

INHALT

Hauptaufsätze

Kleine Mitteilungen

Wir bitten, **Manuskriptsendungen** zweifach (Original und eine Kopie, sprachlich einwandfrei, Formeln mit Maschine oder in Druckschrift geschrieben) an folgende Anschrift zu richten:

Zeitschrift für Angewandte Mathematik und Mechanik z. Hd. Herrn Prof. Dr. Günter Schmidt
Zentralinstitut für Mathematik und Mechanik an der Akademie der Wissenschaften der DDR
DDR-1199 Berlin, Rudower Chaussee 5

Zu den Arbeiten, die als Hauptaufsätze bestimmt sind, ist auf gesondertem Blatt eine Zusammenfassung von 5 bis 10 Zeilen in englischer und (möglichst) deutscher und russischer Sprache beizufügen. Ausführliche Hinweise für die Autoren, um deren strikte Berücksichtigung gebeten wird, finden sich im Anschluß an das Inhaltsverzeichnis des Jahrganges 59 (1979).

Die Autoren erhalten von den Hauptaufsätzen 75, von den Kleinen Mitteilungen 25 Sonderdrucke ohne Berechnung, darüber hinaus weitere Sonderdrucke gegen Berechnung.

Der Verlag behält sich für alle Beiträge das Recht der Vervielfältigung und Übersetzung vor.

Bestellungen sind zu richten

- **in der DDR** an den Postzeitungsvertrieb, an eine Buchhandlung oder an den AKADEMIE-VERLAG, DDR-1080 Berlin, Leipziger Straße 3–4
- **im sozialistischen Ausland** an eine Buchhandlung für fremdsprachige Literatur oder an den zuständigen Postzeitungsvertrieb
- **in der BRD und Westberlin** an eine Buchhandlung oder an die Auslieferungsstelle KUNST UND WISSEN, Erich Bieber, 7000 Stuttgart 1, Wilhelmstraße 4–6
- **in Österreich** an den Globus-Buchvertrieb, 1201 Wien, Höchstädtplatz 3
- **in den übrigen westeuropäischen Ländern** an eine Buchhandlung oder an die Auslieferungsstelle KUNST UND WISSEN, Erich Bieber GmbH, CH-8008 Zürich/Schweiz, Dufourstraße 51
- **im übrigen Ausland** an den Internationalen Buch- und Zeitschriftenhandel; den Buchexport, Volkseigener Außenhandelsbetrieb der Deutschen Demokratischen Republik, DDR-7010 Leipzig, Postfach 160, oder an den AKADEMIE-VERLAG, DDR-1080 Berlin, Leipziger Straße 3–4.

ZEITSCHRIFT FÜR ANGEWANDTE MATHEMATIK UND MECHANIK
Herausgeber und Chefredakteur: Prof. Dr. Günter Schmidt. Redaktion: Dr. Winfried Heinrich, Dr. Horst Weinert, Dipl.-Math. Friedhild Dudel, Helga Rühl, Zentralinstitut für Mathematik und Mechanik der Akademie der Wissenschaften der DDR. Verlag: Akademie-Verlag, DDR-1080 Berlin, Leipziger Straße 3–4; Fernruf 2236221 oder 2236229; Telex-Nr.: 114420; Bank: Staatsbank der DDR, Berlin, Kto.-Nr.: 6836-26-20712. Anschrift der Redaktion: Zentralinstitut für Mathematik und Mechanik der Akademie der Wissenschaften, DDR-1199 Berlin, Rudower Chaussee 5; Fernruf: 6702841/3639.

Veröffentlicht unter der Lizenznummer 1282 des Presseamtes beim Vorsitzenden des Ministerrates der Deutschen Demokratischen Republik. Gesamtherstellung: VEB Druckerei „Thomas Müntzer", DDR-5820 Bad Langensalza. Erscheinungsweise: Die Zeitschrift für Angewandte Mathematik und Mechanik erscheint monatlich. Die 12 Hefte eines Jahres einschließlich Tagungsheft bilden einen Band. Bezugspreis je Band 360,— M zuzüglich Versandspesen (Preis für die DDR 216,— M). Bezugspreis je Heft 30,— M (Preis für die DDR 18,— M). Bestellnummer dieses Heftes: 1009/60/11.

AN (EDV) 35937

ISSN 0044—2267

ZAMM 60, 549 – 555 (1980)

S. K. GUPTA / V. D. RANA

Torsion of a Transversely Isotropic Cylinder in the Theory of Creep

Transitional stresses for a transversely isotropic circular cylinder subjected to torsion have been derived. It has been shown that the asymptotic solution through the stress and stress-difference gives the plastic and creep stresses respectively. The effect of transverse isotropy is presented graphically. The results indicate that the value of maximum shear stress for a cylinder of transversely isotropic material is greater than that of an isotropic incompressible material and it increases with the increase in measure 'n'. For an isotropic incompressible material, as a particular case, the results are the same as given by Marin.

Es wurden die Übergangsspannungen eines transversal isotropen Kreiszylinders bei Torsion bestimmt. Es wurde gezeigt, daß die asymptotische Lösung infolge der Spannung und Spannungsdifferenz entsprechend die plastischen und Kriechspannungen liefert. Der Effekt der transversalen Isotropie wird graphisch veranschaulicht. Die Ergebnisse zeigen, daß der maximale Wert der Schubspannungen bei einem Zylinder mit transversal isotropem Material größer ist als bei einem mit isotrop inkompressiblem Material. Er wächst mit der Vergrößerung des Maßes 'n'. Für den Spezialfall des isotrop inkompressiblen Materials stimmen die Ergebnisse mit denen von Marin überein.

Определены переходные напряжения трансверсально изотропного кругового цилиндра при вращении. Показано, что асимптотическое решение посредством напряжения и разности напряжения даёт пластическое напряжение и напряжение ползучести соответственно. Эффект трансверсальной изотропии представлен графически. Результаты показали, что максимальное значение напряжений сдвига у цилиндра из трансверсально изотропного материала больше, чем у цилиндра из изотропного несжимаемого материала и что это значение возрастает с возрастанием меры ‚n'. Результаты, полученные в частном случае изотропного несжимаемого материала, совпадают с результатами Марина.

Introduction

Creep torsion of circular cylinder for an isotropic material has been discussed by BHATNAGAR [1], MARIN [2], and FINNIE and HELLER [3]; and for an orthotropic material by BHATNAGAR and GUPTA [4]. These authors have analysed the problem after making some simplifying assumptions. Firstly, the deformations are assumed to be small enough to make the infinitesimal strain theory applicable. Secondly, the constitutive equations of the material are simplified by assuming incompressibility of the material together with an yield condition and a creep strain law like that of NORTON. SETH's transition theory [5] does not require any of these assumptions. It utilizes the concept of generalized strain measure and asymptotic transition through the critical points of the differential system defining the deformed field. A number of problems [7, 8, 12, 13] have been solved by using this concept.

For uni-axial case, SETH [6] has defined the generalized strain measure as

$$e = \left[\frac{1}{n}\left\{1 - \left(\frac{l_0}{l}\right)^n\right\}\right]^m \tag{1.1}$$

where n is the measure, m the irreversibility index, l_0 and l are the initial and strained lengths of the rod respectively. For $n = 0, 1, 2, -1, -2$, it gives the HENCKY, SWAINGER, ALMANSI, CAUCHY and GREEN measure respectively.

2. Governing Equations

Consider a circular cylinder of radius a subjected to finite twist. The components of displacement in cylindrical co-ordinates are given by [7, 8],

$$u = r(1 - \beta), \qquad v = \alpha r z, \qquad w = dz, \tag{2.1}$$

where β is a function of $r = (x^2 + y^2)^{1/2}$, α is the angle of twist per unit length and d is a constant.

The generalized components of strain from equation (1.1) are,

$$\begin{aligned} e_{rr} &= \frac{1}{n^m}[1 - (r\beta' + \beta)^n]^m, \qquad e_{\theta\theta} = \frac{1}{n^m}[1 - \beta^n]^m, \\ e_{zz} &= \left(\frac{D^n}{n}\right)^m\left[1 - \left(\frac{\alpha r\beta}{D}\right)^n\right]^m, \qquad e_{\theta z} = \frac{1}{n^m}[\alpha^{n/2} r^{n/2} \beta^n]^m, \end{aligned} \tag{2.2}$$

where, $e_{r\theta} = e_{rz} = 0$,

$$D^n = 1 - (1 - d)^n \quad \text{and} \quad \beta' = \frac{d\beta}{dr}.$$

The stress-strain relations for transversely isotropic materials are [9],

$$\begin{aligned} \tau_{rr} &= c_{11}e_{rr} + (c_{11} - 2c_{66})\, e_{\theta\theta} + c_{13}e_{zz}, \qquad \tau_{\theta\theta} = (c_{11} - 2c_{66})\, e_{rr} + c_{11}e_{\theta\theta} + c_{13}e_{zz}, \\ \tau_{zz} &= c_{13}e_{rr} + c_{13}e_{\theta\theta} + c_{33}e_{zz}, \qquad \tau_{\theta z} = c_{44}e_{\theta z}, \qquad \tau_{r\theta} = \tau_{zr} = 0 \end{aligned} \tag{2.3}$$

where c_{11}, c_{13}, c_{33}, c_{44} and c_{66} are the constants of the material. Using equations (2.2) in equation (2.3), we have

$$\tau_{rr} = \frac{c_{11}}{n^m}[1-(r\beta'+\beta)^n]^m + \frac{c_{11}-2c_{66}}{n^m}[1-\beta^n]^m + \frac{c_{13}}{n^m}(D^m)\left[1-\left(\frac{\alpha r\beta}{D}\right)^n\right]^m,$$

$$\tau_{\theta\theta} = \frac{c_{11}-2c_{66}}{n^m}[1-(r\beta'+\beta)^n]^m + \frac{c_{11}}{n^m}[1-\beta^n]^m + \frac{c_{13}}{n^m}(D^m)\left[1-\left(\frac{\alpha r\beta}{D}\right)^n\right]^m,$$

$$\tau_{zz} = \frac{c_{13}}{n^m}[1-(r\beta'+\beta)^n]^m + \frac{c_{13}}{n^m}[1-\beta^n]^m + \frac{(D^m c_{33})}{n^m}\left[1-\left(\frac{\alpha\beta r}{D}\right)^n\right]^m, \tag{2.4}$$

$$\tau_{\theta z} = \frac{c_{44}}{n^m}[\alpha^{n/2}r^{n/2}\beta^n]^m, \qquad \tau_{r\theta} = \tau_{zr} = 0.$$

Equations of equilibrium are all satisfied except

$$\frac{\partial\tau_{rr}}{\partial r} + \frac{\tau_{rr}-\tau_{\theta\theta}}{r} = 0. \tag{2.5}$$

Using equations (2.4) in equation (2.5), we have a non-linear differential equation satisfied by β as

$$P(P+1)^{n-1}\beta\frac{dP}{d\beta}\{1-\beta^n(P+1)^n\}^{m-1} + P(P+1)^n\{1-\beta^n(P+1)^n\}^{m-1} + \frac{(c_{11}-2c_{66})}{c_{11}}P(1-\beta^n)^{m-1} +$$
$$+ \frac{c_{13}}{c_{11}}(\alpha r)^n(P+1)(D^n)^{m-1}\left\{1-\left(\frac{\alpha r\beta}{D}\right)^n\right\}^{m-1} - \frac{2c_{66}}{c_{11}mn\beta^n}[\{1-\beta^n(P+1)^n\}^m - (1-\beta^n)^m] = 0 \tag{2.6}$$

where

$$r\beta' = \beta P.$$

If $m = 1$, which holds for secondary creep, equation (2.6) reduces to

$$\left[(P+1)^n\left(P+\frac{2c_{66}}{nc_{11}}\right) + \frac{(c_{11}-2c_{66})}{c_{11}}P + (P+1)\frac{c_{13}}{c_{11}}(\alpha r)^n - \frac{2c_{66}}{nc_{11}}\right]\frac{d\beta}{dP} + \beta P(P+1)^{n-1} = 0. \tag{2.7}$$

Equation (2.6) shows that transition points of β are $P = 0$, $P = -1$. The only critical point of interest is $P \to -1$, and asymptotic transition through $P \to -1$ gives the plastic or creep stresses [7, 8, 12, 13] depending upon the transition functions used. The twisting couple M is given by

$$M = 2\pi\int_0^a r^2\tau_{\theta z}\,dr. \tag{2.8}$$

Boundary conditions are

$$\tau_{rr} = 0 \quad\text{at}\quad r = a, \qquad \int_0^a r\tau_{zz}\,dr = 0. \tag{2.9), (2.10}$$

3. Solution through the Stress

In this section it is shown that the asymptotic solution through the stress leads from elastic state to plastic state.

We define the transition function R as

$$R = \frac{2(c_{11}-c_{66})}{c_{11}} + \frac{c_{13}}{c_{11}}D^n - n\frac{\tau_{rr}}{c_{11}}. \tag{3.1}$$

Substituting the value of τ_{rr} from equation (2.4) in equation (3.1), we have for $m = 1$,

$$R = \frac{2(c_{11}-c_{66})}{c_{11}} + \frac{c_{13}}{c_{11}}D^n - \frac{1}{c_{11}}\left[c_{11}\{1-\beta^n(P+1)^n\} + (c_{11}-2c_{66})(1-\beta^n) + c_{13}D^n\left\{1-\left(\frac{\alpha r\beta}{D}\right)^n\right\}\right]. \tag{3.2}$$

Taking logarithmic differentiation of equation (3.2) with respect to r, we get

$$\frac{d(\log R)}{dr} = \frac{n\beta^n}{rRc_{11}}\left[c_{11}P(P+1)^{n-1}\beta\frac{dP}{d\beta} + P\{(c_{11}-2c_{66}) + c_{13}\alpha^n r^n\} + c_{13}\alpha^n r^n + c_{11}P(P+1)^n\right]. \tag{3.3}$$

Substituting the value of $dP/d\beta$ from equation (2.7) in equation (3.3) and taking the asymptotic value $P \to -1$, we get

$$\frac{d}{dr}(\log R) = \frac{2c_{66}}{r\{(c_{11}-2c_{66}) + c_{13}\alpha^n r^n\}}. \tag{3.4}$$

Integrating equation (3.4), we get

$$R = K_1\left[\frac{r^n}{1+K\alpha^n r^n}\right]^{2c_{66}/[n(c_{11}-2c_{66})]} \tag{3.5}$$

where K_1 is a constant of integration and $K = c_{13}/(c_{11}-2c_{66})$.

The asymptotic value of β is obtained from equation (3.1) and (3.5) as

$$\beta^n = \frac{K_1 c_{11}}{(c_{11} - 2c_{66})(1 + K\alpha^n r^n)} \left[\frac{r^n}{1 + K\alpha^n r^n}\right]^{2c_{66}/[n(c_{11} - 2c_{66})]} \tag{3.6}$$

Using equation (3.5) in equation (3.1), we get

$$\tau_{rr} = \frac{c_{11}}{n}\left[\frac{2(c_{11} - c_{66})}{c_{11}} + \frac{c_{13}}{c_{11}} D^n - K_1 \left\{\frac{r^n}{1 + K\alpha^n r^n}\right\}^{2c_{66}/[n(c_{11} - 2c_{66})]}\right]. \tag{3.7}$$

Putting the value of τ_{rr} from equation (3.7) in equation (2.5), we have

$$\tau_{\theta\theta} = \frac{c_{11}}{n}\left[\frac{2(c_{11} - c_{66})}{c_{11}} + \frac{c_{13}}{c_{11}} D^n - K_1 \left\{\frac{r^n}{1 + K\alpha^n r^n}\right\}^{2c_{66}/n(c_{11} - 2c_{66})} \frac{(c_{11} - 2c_{66}) K\alpha^n r^n + c_{11}}{(c_{11} - 2c_{66})(1 + K\alpha^n r^n)}\right]. \tag{3.8}$$

The asymptotic value of τ_{zz} for $m = 1$, is obtained from equations (2.4) and (3.6) as

$$\tau_{zz} = \frac{1}{n}\left[2c_{13} + c_{33}D^n - \frac{c_{11}K_1}{(c_{11} - 2c_{66})}\left\{\frac{c_{13} + c_{33}\alpha^n r^n}{1 + K\alpha^n r^n}\right\}\left\{\frac{r^n}{1 + K\alpha^n r^n}\right\}^{2c_{66}/n(c_{11} - 2c_{66})}\right]. \tag{3.9}$$

Applying boundary conditions (2.9) and (2.10) in equations (3.7) and (3.9) respectively, we get

$$K_1 = \left[\frac{2(c_{11} - c_{66})}{c_{11}} + \frac{c_{13}}{c_{11}} D^n\right]\left[\frac{a^n}{1 + K\alpha^n a^n}\right]^{-2c_{66}/n(c_{11} - 2c_{66})} \tag{3.10}$$

and

$$D^n = -\frac{2c_{13}}{c_{33}} + \\ + \frac{2K_1 c_{11}}{c_{33}(c_{11} - 2c_{66})}\left[(c_{13} + c_{33}\alpha^n)\left\{\frac{a^{2c_{66}/(c_{11} - 2c_{66})}}{2(c_{11} - c_{66})}(c_{11} - 2c_{66}) + \frac{[2c_{66}(n-1) - nc_{11}]\, K\alpha^n a^{2c_{66}(1-n) + nc_{11}}}{n[c_{11}(n+2) - 2c_{66}(n+1)]} + \dots\right\}\right]. \tag{3.11}$$

The value of the material constant in the transition range is given by [10],

$$Y = \frac{4c_{66}}{n} \cdot \frac{[c_{33}(c_{11} - c_{66}) - c_{13}^2]}{(c_{11}c_{33} - c_{13}^2)} \tag{3.12}$$

where Y is the yield stress in tension.

The corresponding values of the stresses are given by

$$\begin{aligned} \tau_{rr} &= \frac{c_{11} Y (c_{11}c_{33} - c_{13}^2)}{4c_{66}[c_{33}(c_{11} - c_{66}) - c_{13}^2]}\left[\frac{2(c_{11} - c_{66})}{c_{11}} + \frac{c_{13}D^n}{c_{11}} - K_1\left\{\frac{r^n}{1 + K\alpha^n r^n}\right\}^{2c_{66}/n(c_{11} - 2c_{66})}\right], \\ \tau_{\theta\theta} &= \frac{c_{11} Y (c_{11}c_{33} - c_{13}^2)}{4c_{66}[c_{33}(c_{11} - c_{66}) - c_{13}^2]}\left[\frac{2(c_{11} - c_{66})}{c_{11}} + \right. \\ &\left. + \frac{c_{13}}{c_{11}} D^n - K_1\left\{\frac{r^n}{1 + K\alpha^n r^n}\right\}^{2c_{66}/n(c_{11} - 2c_{66})} \frac{(c_{11} - 2c_{66}) K\alpha^n r^n + c_{11}}{(c_{11} - 2c_{66})(1 + K\alpha^n r^n)}\right], \\ \tau_{zz} &= \tau_{rr} + \frac{1}{n}[2(c_{13} - c_{11} + c_{66}) - \beta^n(c_{13} - c_{11} + 2c_{66}) + (c_{33} - c_{13})\{D^n - (\alpha r\beta)^n\}] \end{aligned} \tag{3.13}$$

where D and K_1 are given by equations (3.10) and (3.11). The shearing stress $\tau_{\theta z}$ is given by equations (2.4) and (3.6) as

$$\tau_{\theta z} = \frac{c_{11}c_{44}K_1\alpha^{n/2} r^{n/2 + 2c_{66}/(c_{11} - 2c_{66})}}{n(c_{11} - 2c_{66})}[1 + K\alpha^n r^n]^{-A} \tag{3.14}$$

where

$$A = \frac{2c_{66}}{n(c_{11} - 2c_{66})} + 1\,.$$

The maximum shearing stress occurs at $r = a$, where plasticity first sets in. Denoting its greatest value by τ, we have

$$\tau = \frac{c_{11}c_{44}K_1\alpha^{n/2} a^{(n/2) + 2c_{66}/(c_{11} - 2c_{66})}}{n(c_{11} - 2c_{66})}[1 + K\alpha^n a^n]^{-A}\,. \tag{3.15}$$

The twisting couple M is given by

$$M = 2\pi\int_0^a r^2\tau_{\theta z}\,dr = \frac{2\pi c_{11}c_{44}K\alpha^{n/2}a^{B+1}}{n(c_{11} - 2c_{66})}\left[\frac{1}{B+1} - \frac{Ak\alpha^n a^n}{B+n+1} + \dots\right] \tag{3.16}$$

where

$$B = 2 + \frac{n}{2} + \frac{2c_{66}}{c_{11} - 2c_{66}}.$$

Eliminating c_{44} between equations (3.15) and (3.16), we have for the first approximation,

$$M = \frac{2\pi a^3 \tau}{3 + \frac{n}{2} + \frac{2c_{66}}{c_{11} - 2c_{66}}}. \tag{3.17}$$

For fully plastic case [10], $c_{66} \to 0$, equations (3.13)—(3.17) become

$$\begin{aligned} \tau_{rr} &= \frac{Y(2c_{11} + c_{13}D^n)}{2nc_{11}}\left[\log\left\{\left(\frac{a}{r}\right)^n \frac{1 + K\alpha^n r^n}{1 + K\alpha^n a^n}\right\}\right], \\ \tau_{\theta\theta} &= \frac{Y(2c_{11} + c_{13}D^n)}{2nc_{11}}\left[\log\left\{\left(\frac{a}{r}\right)^n \frac{1 + K\alpha^n r^n}{1 + K\alpha^n a^n}\right\} - \frac{n}{1 + K\alpha^n r^n}\right], \\ \tau_{zz} &= \tau_{rr} + \frac{1}{n}\left[2(c_{13} - c_{11}) - \beta^n(c_{13} - c_{11}) + (c_{33} - c_{13})\left\{D^n - (\alpha r\beta)^n\right\}\right], \\ \tau_{\theta z} &= \frac{c_{44}\alpha^{n/2} r^{n/2} \cdot (2c_{11} + c_{13}D^n)}{nc_{11}(1 + K\alpha^n r^n)}, \end{aligned} \tag{3.18}$$

and

$$M = \frac{2\pi a^3 \tau}{3 + \frac{n}{2}}.$$

Particular Case

For isotropic materials, the constants reduce to two only [11] i.e. $c_{11} = c_{33}$, $c_{13} = c_{11} - 2c_{66}$, $c_{44} = c_{66}$. Equations (3.13)—(3.17) reduce to

$$\begin{aligned} \tau_{rr} &= \frac{Y(2 - c)}{c(3 - 2c)}\left[(3 - 2c) - (1 - c)(1 - D^n) - K_1\left\{\frac{r^n}{1 + \alpha^n r^n}\right\}^{c/n(1-c)}\right], \\ \tau_{\theta\theta} &= \frac{Y(2 - c)}{c(3 - 2c)}\left[(3 - 2c) - (1 - c)(1 - D^n) - K_1\left\{\frac{r^n}{1 + \alpha^n r^n}\right\}^{c/n(1-c)} \frac{(1 - c)\alpha^n r^n + 1}{(1 - c)(1 + \alpha^n r^n)}\right], \\ \tau_{zz} &= \tau_{rr} + \frac{Y(2 - c)}{(3 - 2c)}\left[D^n - (\alpha r\beta)^n - 1\right], \\ \tau_{\theta z} &= \frac{Y(2 - c)\, K_1 \alpha^{n/2} r^{(n/2) + c/(1-c)}}{2(3 - 2c)(1 - c)}\left[1 + \alpha^n r^n\right]^{-A}, \\ M &= \frac{2\pi a^3 \tau}{3 + (n/2) + c(1 - c)}. \end{aligned} \tag{3.19}$$

where

$$Y = \frac{2\mu(3 - 2c)}{n(2 - c)} \quad \text{and} \quad \frac{2c_{66}}{c_{11}} = c.$$

If we take $n = 2$, the equations (3.19) for fully plastic case (i.e. $c \to 0$) become,

$$\begin{aligned} \tau_{rr} &= \frac{Y}{3}\left[(2 + D^2)\log\left\{\left(\frac{a}{r}\right)^2 \frac{1 + \alpha^2 r^2}{1 + \alpha^2 a^2}\right\}\right], \quad \tau_{\theta\theta} = \frac{Y}{3}\left[(2 + D^2)\log\left\{\left(\frac{a}{r}\right)^2 \frac{1 + \alpha^2 r^2}{1 + \alpha^2 a^2}\right\} - \frac{2}{1 + \alpha^2 r^2}\right], \\ \tau_{zz} &= \tau_{rr} + \frac{2Y}{3}\left[D^2 - (\alpha r\beta)^2 - 1\right], \quad \tau_{\theta z} = \frac{Y}{3}\alpha r\left[\frac{2 + D^2}{1 + \alpha^2 r^2}\right], \end{aligned} \tag{3.20}$$

and

$$M = \tfrac{1}{2}\pi a^3 \tau$$

where

$$D^2 = 2\left[\frac{\alpha^2 a^2}{\log(1 + \alpha^2 a^2)} - 1\right].$$

These equations are the same as obtained by SETH [7] for elastic-plastic transition in torsion. For other values of measure n, it gives creep transition, whose discussion will follow in section 4.

4. Solution through the Stress Difference

It has been shown [8, 12, 13] that the transition function through the stress-difference at the transition point $P \to -1$ gives the creep stresses. The transition function is taken as

$$R_2 = \tau_{rr} - \tau_{\theta\theta} = \frac{2c_{66}}{n^m} [\{1 - (r\beta' + \beta)^n\}^m - (1 - \beta^n)^m] . \tag{4.1}$$

Taking the logarithmic differentiation of equation (4.1) with respect to β, we get

$$\frac{d}{d\beta}(\log R_2) = \frac{mn\beta^{n-1}\left[(1-\beta^n)^{m-1} - \{1-\beta^n(P+1)^n\}^{m-1}\{(P+1)^n + (P+1)\}^{n-1}\cdot \beta \dfrac{dP}{d\beta}\right]}{\{1-\beta^n(P+1)^n\}^m - (1-\beta^n)^m} . \tag{4.2}$$

Using equation (2.6) in equation (4.2) and taking $P \to -1$, we get

$$\frac{d}{d\beta}(\log R_2) = \frac{2c_{66}}{c_{11}\beta} + \frac{2mn\beta^{n-1}(1-\beta^n)^{m-1}}{1-(1-\beta^n)^m}\,\frac{(c_{11}-c_{66})}{c_{11}} . \tag{4.3}$$

Integrating equation (4.3), we have

$$R_2 = Ar^{-2c_{66}/c_{11}}[1 - (1-\beta^n)^m]^{2(c_{11}-c_{66})/c_{11}} \tag{4.4}$$

where A is a constant of integration.

For secondary stage of creep (i.e. for $m = 1$) equation (4.4) becomes

$$R_2 = \tau_{rr} - \tau_{\theta\theta} = A_1 r^{-2n+2c_{66}(n-1)/c_{11}} ; \tag{4.5}$$

Substituting equation (4.5) in equation (2.4) and integrating we get

$$\tau_{rr} = \frac{A_1 r^{-2n+2c_{66}(n-1)/c_{11}}}{2n - (2c_{66}(n-1)/c_{11})} + B \tag{4.6}$$

where B is a constant of integration.

The asymptotic values of other stresses can be written from equations (4.5), (2.2) and (2.3) as

$$\tau_{\theta\theta} = A_1 r^{-2n+(2c_{66}(n-1)/c_{11})}\left[\frac{1}{2n-(2c_{66}(n-1)/c_{11})} - 1\right] + B , \tag{4.7}$$

$$\tau_{zz} = \frac{c_{33}}{2c_{13}} A_1 r^{-2n+(2c_{66}(n-1)/c_{11})}\left[\frac{2}{2n-(2c_{66}(n-1)/c_{11})} - 1\right] + \frac{B(c_{33})}{c_{13}} + \frac{c_{13}^2 - (c_{11}-c_{66})\,c_{33}}{nc_{13}}(2-\beta^n) . \tag{4.8}$$

The constants A_1 and B have to be determined from boundary conditions (2.9) and (2.10). Using (2.9) in equation (4.6), we get

$$B = \frac{-A_1 a^{-2n+(2c_{66}(n-1)/c_{11})}}{2n - (2c_{66}(n-1)/c_{11})} . \tag{4.9}$$

Putting the value of constant B in equations (4.6)—(4.8), we get the transitional creep stresses as

$$\begin{aligned}
\tau_{rr} &= \frac{A_1 a^{-2n+(2c_{66}(n-1)/c_{11})}}{2n-(2c_{66}(n-1)/c_{11})}\left[\left(\frac{r}{a}\right)^{-2n+(2c_{66}(n-1)/c_{11})} - 1\right],\\
\tau_{\theta\theta} &= \frac{A_1 a^{-2n+(2c_{66}(n-1)/c_{11})}}{2n-(2c_{66}(n-1)/c_{11})}\left[\left\{1 - 2n + \frac{2c_{66}}{c_{11}}(n-1)\right\}\left(\frac{r}{a}\right)^{-2n+(2c_{66}(n-1)/c_{11})} - 1\right],\\
\tau_{zz} &= \frac{c_{33}}{2c_{13}}\,\frac{A_1 a^{-2n+(2c_{66}(n-1)/c_{11})}}{2n-(2c_{66}(n-1)/c_{11})}\left[\left\{2 - 2n + \frac{2c_{66}}{c_{11}}(n-1)\right\}\left(\frac{r}{a}\right)^{-2n+(2c_{66}(n-1)/c_{11})} - 2\right] +\\
&\quad + \frac{c_{13}^2 - (c_{11}-c_{66})\,c_{33}}{nc_{13}}(2-\beta^n)
\end{aligned} \tag{4.10}$$

where the value of A_1 is obtained from equations (2.10) and (4.10) as

$$A_1 = \frac{2c_{66}}{n}\left[2 - 2n + \frac{2c_{66}}{c_{11}}(n-1)\right] a^{2n-(2c_{66}(n-1)/c_{11})} .$$

The asymptotic value of β from equations (4.1) and (4.4) is

$$\beta^n = \left(\frac{2c_{66}}{nA}\right)^{c_{11}/(c_{11}-2c_{66})} r^{2c_{66}/(c_{11}-2c_{66})} . \tag{4.11}$$

Using equation (4.11) in equation (2.3), the shearing stress is given by

$$\tau_{\theta z} = \frac{c_{44}}{n}\alpha^{n/2}\left(\frac{2c_{66}}{nA}\right)^{c_{11}/(c_{11}-2c_{66})} r^{(n/2)+2c_{66}/(c_{11}-2c_{66})} . \tag{4.12}$$

The twisting couple M is

$$M = 2\pi \int_0^a r^2 \tau_{\theta z}\, dr = \frac{2\pi c_{44}}{n} \alpha^{n/2} \left(\frac{2c_{66}}{nA}\right)^{c_{11}/(c_{11}-2c_{66})} a^{(2c_{66}/(c_{11}-2c_{66}))+(n/2)+3} \Big/ \left(3 + \frac{n}{2} + \frac{2c_{66}}{c_{11}-2c_{66}}\right). \tag{4.13}$$

Eliminating c_{44} between equations (4.12) and (4.13), we get

$$\tau_{\theta z} = \frac{1}{4}\left[3 + \frac{n}{2} + \frac{2c_{66}}{c_{11}-2c_{66}}\right]\left(\frac{r}{a}\right)^{(n/2)+2c_{66}/(c_{11}-2c_{66})} \frac{2M}{\pi a^3}. \tag{4.14}$$

The maximum shear stress occurs at $r = a$, denoting its value by τ, we have

$$\tau = \frac{1}{4}\left[3 + \frac{n}{2} + \frac{2c_{66}}{c_{11}-2c_{66}}\right]\tau_e \tag{4.15}$$

where τ_e is the elastic maximum shear stress equal to $2M/\pi a^3$.

This expression for maximum shearing stress is the same as equation (3.17) which is obtained through the transition function R.

5. Isotropic Steady State of Creep

For isotropic materials, the constants reduce to two only [11] i.e., $c_{11} = c_{33}$, $c_{13} = c_{11} - 2c_{66}$, $c_{44} = c_{66}$. Equation (4.5) then becomes

$$\tau_{rr} - \tau_{\theta\theta} = A_1 r^{-2n+c(n-1)} \tag{5.1}$$

where

$$\frac{2c_{66}}{c_{11}} = c\,.$$

For HENCKY measure ($n \to 0$) which holds good for stationary state of creep, equation (5.1) reduces to the condition of TRESCA type for incompressible material ($c \to 0$) i.e.,

$$\tau_{rr} - \tau_{\theta\theta} = A_1 \equiv \sigma_T\,.$$

where σ_T is the yield stress in tension.

Creep transitional stresses (4.10) for isotropic material become,

$$\begin{aligned}
\tau_{rr} &= \frac{A_1 a^{-2n+c(n-1)}}{2n - c(n-1)}\left[\left(\frac{r}{a}\right)^{-2n+c(n-1)} - 1\right],\\
\tau_{\theta\theta} &= \frac{A_1 a^{-2n+c(n-1)}}{2n - c(n-1)}\left[\{1 - 2n + c(n-1)\}\left(\frac{r}{a}\right)^{-2n+c(n-1)} - 1\right],\\
\tau_{zz} &= \frac{A_1}{2-c}\,\frac{a^{-2n+c(n-1)}}{2n - c(n-1)}\left[\{2 - 2n + c(n-1)\}\left(\frac{r}{a}\right)^{-2n+c(n-1)} - 2\right] + \frac{(3-2c)}{(1-c)}\frac{c_{66}}{n}(2 - \beta^n)
\end{aligned} \tag{5.2}$$

where

$$A_1 = \frac{2c_{66}}{n}[2 - 2n + c(n-1)]\, a^{2n-c(n-1)}$$

and shearing stress from equation (4.15) reduces to

$$\tau_{\theta z} = \frac{1}{4}\left[3 + \frac{n}{2} + \frac{c}{1-c}\right]\left(\frac{r}{a}\right)^{(n/2)+c/(1-c)} \frac{2M}{\pi a^3}. \tag{5.2}$$

Maximum shear stress is given by

$$\tau = \frac{1}{4}\left[3 + \frac{n}{2} + \frac{c}{1-c}\right]\tau_e \tag{5.3}$$

where τ_e is the elastic maximum shear stress.

For incompressible material i. e. $c \to 0$, equation (5.3) becomes

$$\tau = \frac{1}{4}\left[\frac{6+n}{2}\right]\tau_c\,. \tag{5.4}$$

This expression for maximum shear stress is the same as given by MARIN [2], provided we put $n = 2/N$.

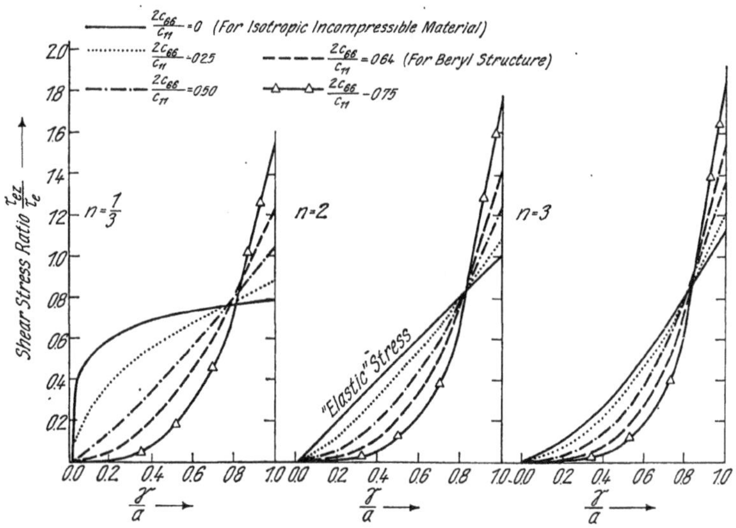

Fig. 1. "CREEP" stress distribution for transversely isotropic cylinder under torsion

6. Numerical Illustration

Curves have been drawn in Fig. 1 between shear stress ratio and the relative distance from the centre for different ratios of $2c_{66}/c_{11}$ and n. $2c_{66}/c_{11} = 0$ represents the curve for isotropic incompressible material and $2c_{66}/c_{11} = 0.64$ for Beryl material [9]. The value $n = 1/3$ (i.e. $N = 6$) is considered to take account of the classical results obtained by MARIN.

These curves illustrate the effect of transverse isotropy of the material and n on shear stress. The curves show that the shear stress is maximum at the outer surface and increases with the increase in anisotropy of the material.

For $n = 2$ and $(2c_{66}/c_{11}) = 0$, elastic shear stress curve is obtained and for other values of $2c_{66}/c_{11}$ and $n = 2$, the resulting curves give the transverse isotropic effects for elastic shear stress distribution.

The results indicate that the value of maximum shear stress ratio for a cylinder of transversely isotropic material is greater than that of an isotropic incompressible material. Further the shear stress is maximum for transversely isotropic material and increases with the increase in measure n. The curve drawn for $n = \frac{1}{3}$ and $(2c_{66}/c_{11}) = 0$ is the same as given by MARIN [2].

Acknowledgement

One of the authors (V. D. RANA) gratefully acknowledges the University Grants Commission for financial assistance.

References

1 BHATNAGAR, N. S., Torsion of solid cylindrical rod of circular cross-section in the theory of creep, Research Journal University of Roorkee, India (1962).
2 MARIN, J., Mechanical Behaviour of Engineering Materials, Prentice-Hall, Inc., 385—387 (1966).
3 FINNIE, I., and HELLER, W. R., Creep of Engineering Materials, McGraw Hill Book Co., New York (1959).
4 BHATNAGAR, N. S., GUPTA, S. K., and GUPTA, R. P., Torsion of an orthotropic rod in the theory of creep, Wood Sci. and Technology 3, 167—174 (1969).
5 SETH, B. R., Transition theory of elastic-plastic deformation, creep and relaxation, Nature, 195, 896—897 (1962).
6 SETH, B. R., Measure concept in Mechanics, Int. Jr. Non-Linear Mechanics, 1, 35—40 (1966).
7 SETH, B. R., Elastic-plastic transition in torsion, ZAMM 44, 229—233 (1964).
8 SETH, B. R., Transition analysis of collapse of thick cylinders, ZAMM 50, 617—621 (1970).
9 LOVE, A. E. H., A Treatise on the Mathematical Theory of Elasticity, 161—163 (Fourth Edition) (1944).
10 SETH, B. R., Transition conditions, The yield Condition. Int. Jr. Non-Linear Mechanics 5, 279—285 (1970).
11 SOKOLNIKOFF, I. S., Mathematical Theory of Elasticity, McGraw Hill Book Co. (Second Edition), 66, (1950).
12 HULSARKAR, S., Transition theory of shells under uniform pressure, ZAMM 46, 431—437 (1966).
13 GUPTA, S. K., and DHARMANI, R. L., Creep transition in transversely isotropic shells under uniform pressure, Indian Jr. of Pure and Applied Maths 8, No. 9, 1049—1054 (1977).

Eingereicht am 7. 5. 1979

Anschrift: Dr. S. K. GUPTA, Department of Mathematics, Himachal Pradesh University, Simla — 171005, India

ZAMM 60, 557 – 567 (1980)

B. SCHULTRICH / M. POSSELT

Statistische Rißausbreitung in eindimensionalen Systemen ohne Wechselwirkung

Die Bruchstatistik berücksichtigt den Einfluß der lokalen Festigkeitsschwankungen auf die Entwicklung der Vorschädigung und deren Übergang in den makroskopischen Bruch. Die hierzu gewöhnlich benutzte Einzelrißnäherung wird an einem eindimensionalen Modellsystem aus nicht miteinander wechselwirkenden Elementen getestet. Für die offene und die geschlossene Elementekette werden Mittelwert und Schwankung der Rißlängenverteilung sowie die makroskopische Bruchwahrscheinlichkeit berechnet. Die Näherungsbeziehungen lassen sich hieraus als asymptotischer Grenzfall für große Systeme gewinnen. Die Rolle der Randeffekte und der statistischen Korrelationen, in deren Vernachlässigung die Einzelrißnäherung besteht, wird genauer untersucht.

The failure statistics considers the influence of local strength fluctuations on the development of precritical cracks and on macroscopic failure behaviour. The single-crack approximation usually used is tested by means of a one-dimensional modelsystem composed of non-interacting elements. Average and variance of the crack-length distribution and the macroscopic fracture probability are calculated. The approximative expressions mentioned above are derived immediately in the limit of large systems. The influence of boundary effects and statistical correlations — neglected in the single-crack approximation — are thoroughly investigated.

Статистика трещин учитывает влияние локальных флуктуаций прочности на развитие докритических дефектов и на их переход в макроскопическое разрушение. Обычно использованная аппроксимация для одной трещины проверяется с помощью одномерной системы невзаимодействующих элементов. Расчитанны среднее значение и среднее квадратичное значение распределения длины трещины, а также вероятность макроскопического разрушения. Приближенные отношения аппроксимации получаются в виде асимптотик для случая больших систем. Исследуется влияние граничных эффектов и статистических корреляций которые пренебрегаются при аппроксимации для одной трещины.

1. Modelle der Bruchstatistik

In zunehmendem Maße werden als Konstruktionswerkstoffe heterogene Materialien mit hochfesten Komponenten eingesetzt. Deren hohe Festigkeit ist im allgemeinen mit einem spröden Verformungsverhalten gekoppelt. Die nur geringfügigen plastischen Deformationen können dann Spannungskonzentrationen nicht wesentlich abbauen, so daß Defekte im Innern und an der Oberfläche den Bruchverlauf stark beeinflussen. Die statistische Verteilung der Schwachstellen in den tragenden Elementen bedingt so notwendig Festigkeitsschwankungen. Die Bruchstatistik, ein Zweig der angewandten Wahrscheinlichkeitstheorie, untersucht den Einfluß der Festigkeitsschwankungen der Elemente auf die makroskopischen Eigenschaften des Festkörpers. Die klassischen Modelle der Bruchstatistik — das Modell des schwächsten Kettengliedes und das Bündelmodell — erfassen die Grenzfälle des Stabilitätsverhaltens. Beim Modell des schwächsten Kettengliedes ist die Wechselwirkung mit den Nachbarelementen so stark, daß ein einmal gebildeter Riß nicht wieder stabilisiert werden kann und sich zur makroskopischen Bruchfläche erweitert [1, 2, 3, 4, 5]. Beim Bündelmodell werden dagegen die unmittelbaren Wechselwirkungen mit dem umgebenden Volumen vernachlässigt: Ein Einzelriß wird vollständig stabilisiert und kann nur über die gleichmäßige Umverteilung der Last auf alle noch tragenden Komponenten eine Ausbreitung der Schädigung bewirken [1, 6, 7, 8]. In vielen Fällen muß jedoch die sich unter einer äußeren Belastung einstellende Schädigungsstruktur genauer berücksichtigt werden. Die hierzu entwickelten Modelle lassen sich auf die folgenden gemeinsamen Grundannahmen zurückführen [9]:

1. Der Körper kann in eine Vielzahl von n ($\gg 1$) Elementen zerlegt werden (z. B. die Kristallkörner, einzelne Faserabschnitte, eingelagerte Teilchen mit ihrer Umgebung usw.).
2. Innerhalb eines Elementes ist keine Rißstabilisierung möglich: Es ist entweder gebrochen oder nicht gebrochen. Der Einfluß verschiedener Bruchmoden wird im einzelnen Element nicht betrachtet.
3. Das Element bricht, sobald die Elementespannung σ_k die Elementebruchspannung σ_{Bk} übersteigt: $\sigma_k \geqq \sigma_{Bk}$. Die im betrachteten Element herrschende Spannung σ_k hängt von der äußeren Belastung σ und der Systemstruktur, d. h. dem Bruchzustand der übrigen Elemente, ab. Die Elementebruchspannung σ_{Bk} beschreibt die innere Defektstruktur des Elementes.
4. Die für ein Element möglichen Werte der Elementebruchspannungen werden durch eine Wahrscheinlichkeitsverteilung $F_k(\sigma') = P(\sigma_{Bk} \leqq \sigma')$ charakterisiert. Sie gibt die Wahrscheinlichkeit an, daß σ_{Bk} unterhalb einer vorgegebenen Spannung σ' liegt oder — mit anderen Worten — daß das Element bei einer lokalen Belastung σ' bricht. Die Elementebruchspannungen verschiedener Elemente sind statistisch unabhängig.

An die Stelle oder an die Seite der Festigkeitsschwankungen können auch Fluktuationen der inneren Spannungsfelder [4, 9, 10] treten. Sie können jedoch zu einer effektiven Bruchwahrscheinlichkeit der Elemente zusammengefaßt und so auf den obigen Fall zurückgeführt werden. Dagegen sollen die kinetischen Aspekte der Rißausbreitung aus dem Problemkreis der vorliegenden Arbeit ausgeklammert bleiben, sie behandelt also das statistische Problem der Kurzzeitfestigkeit. Eine zusammenfassende Darstellung der atomistischen [11, 12, 13, 14, 15], der mikroskopischen [16, 17] und der mesoskopischen Theorien [18, 19] der zeitabhängigen Schädigungsentwicklung wird in [20] gegeben. Unberücksichtigt bleibt auch der ganz ähnlich gelagerte Fall einer Wechselbelastung (bei der die Zyklenzahl die Rolle des Ablaufparameters übernimmt) sowie der Einfluß stochastischer äußerer Kräfte [4].

Die verschiedenen Modelle der Bruchstatistik unterscheiden sich einmal in dem Grad, in dem die schädigungsbedingten Spannungsumverteilungen berücksichtigt werden. Die volle Rückwirkung der Elementeausfälle auf ein noch nicht gebrochenes Element läßt sich im Rahmen eines allgemeinen Gleichungssystems für die Strukturwahrscheinlichkeit des Gesamtsystems beschreiben [9, 21]. Deren Auswertung ist jedoch für makroskopische Systeme (mit $n \gg 1$) — von einfachen Spezialfällen wie dem Bündelmodell abgesehen — nur mit Hilfe zusätzlicher Näherungsannahmen möglich. Gewöhnlich beschränkt man sich deshalb auf die Beeinflussung der unmittelbar angrenzenden Elemente [22, 23, 24], oder es werden die Spannungsumverteilungen vollständig vernachlässigt [25, 26, 27].

Bei einer unvollständigen Berücksichtigung der strukturabhängigen Spannungsüberhöhungen folgt die Systeminstabilität, d. h. der makroskopische Bruch, nicht mehr zwangsläufig aus der Schädigungsakkumulation, sondern muß zusätzlich über die Angabe von Grenzzuständen postuliert werden. Wenn die spröden, festen Einlagerungen voneinander isoliert sind und nur einen geringen Teil des Gesamtkörpers ausmachen, so kann der Bruch durch das Überschreiten einer kritischen Rißdichte ausgelöst werden [10]. Bei dominierenden Sprödphasen wird dagegen die Bildung von Makrorissen, d. h. von zusammenhängenden Komplexen aus gebrochenen Elementen, zum festigkeitsbestimmenden Prozeß. In Anlehnung an die GRIFFITHsche Theorie des determinierten Sprödbruches kann dann die Herausbildung eines Risses von überkritischer Größe als Bruchkriterium benutzt werden [22, 23, 24, 25] (In [26,28] wurde die kritische Rißgröße mit der zerteilenden Bruchfläche selbst identifiziert).

Bewußt oder unbewußt wird bei der statistischen Behandlung der Makrorisse die Einzelrißnäherung angewandt, d. h. es wird die Wechselwirkung zwischen den verschiedenen Rissen weitgehend vernachlässigt. Alle im Körper überhaupt möglichen Risse lassen sich gemäß ihrer Lage eindeutig den n Elementen zuordnen. Die dem gleichen Element zugehörigen Risse bezeichnen wir als eine Rißgruppe. Form und Größe des einzelnen Risses werden durch einen Parametersatz (N) erfaßt. Im allgemeinen begnügt man sich mit der Angabe der Rißgröße N, d. h. der Anzahl der im Riß enthaltenen gebrochenen Elemente. Als die zentralen statistischen Größen in der Einzelrißnäherung erweisen sich die Rißwahrscheinlichkeiten $p_N(\sigma)$, die Wahrscheinlichkeiten, daß von den möglichen Rissen einer bestimmten Rißgruppe gerade ein N-Riß verwirklicht wird. In eindimensionalen Elementeanordnungen gibt es in jeder Rißgruppe nur jeweils einen Riß einer bestimmten Größe und $p_N(\sigma)$ fällt mit der Realisierungswahrscheinlichkeit eines herausgegriffenen N-Risses zusammen. In großen Systemen $(n \gg 1)$ mit kleinen Rissen $(N \ll n)$ sind physikalische Wechselwirkungen (über die Spannungsfelder) und geometrische Wechselwirkungen (durch Rißvereinigungen) zwischen Rissen, die zu verschiedenen Rißgruppen gehören, ohne große Bedeutung. Nach [21] lassen sich dann die effektiven und die Brucheigenschaften auf die Rißwahrscheinlichkeiten $p_N(\sigma)$ zurückführen. Globale Eigenschaften wie der effektive Elastizitätsmodul hängen vom Strukturzustand des ganzen Körpers ab. Er kann (wieder bei Vernachlässigung der Rißwechselwirkung) durch das Rißspektrum ν_N, die Zahl $q_N(n)$ der verschiedenen N-Risse bezogen auf die Gesamtzahl n der Elemente, charakterisiert werden. Das Rißspektrum wiederum läßt sich näherungsweise durch die Rißwahrscheinlichkeiten $p_N(\sigma)$ ausdrücken [21]:

$$\nu_N \equiv \frac{q_N(n)}{n} \approx \frac{\overline{q_N(n)}}{n} \approx p_N(\sigma) \tag{1}$$

(die Mittelung bezieht sich auf eine Schar von Proben mit verschiedenen Elementebruchspannungen). Der makroskopische Bruch wird im allgemeinen durch den größten vorhandenen Riß ausgelöst: Der Körper bricht, sobald sich wenigstens ein Riß ausbildet, der eine gewisse (unter Umständen von der äußeren Belastung abhängige) kritische Größe N_c übersteigt. Die makroskopische Bruchwahrscheinlichkeit $F_A = F_{N_c}$ kann nach [9] näherungsweise durch die Wahrscheinlichkeiten Q_{N_c} ausgedrückt werden, daß sich an einem Element ein instabiler Riß $N \geqq N_c$ entwickelt:

$$F_A = F_{N_c} \approx 1 - (1 - Q_{N_c})^n \approx 1 - \mathrm{e}^{-nQ_{N_c}}, \tag{2}$$

$$Q_{N_c} = 1 - \sum_{N=0}^{N_c-1} p_N. \tag{3}$$

Die für $n \gg 1$, $N_c \ll n$ gültigen Gleichungen (1) bis (3) spielen eine zentrale Rolle in der Bruchstatistik, da durch sie das statistische Problem auf die Berechnung der Rißwahrscheinlichkeit p_N reduziert wird.

2. Eindimensionale Systeme ohne Wechselwirkung

In der vorliegenden Arbeit sollen für ein Modellsystem das Rißspektrum ν_N und die makroskopische Bruchwahrscheinlichkeit F_A für beliebige n, N und N_c berechnet und die asymptotischen Beziehungen (1) und (2) als Grenzfall $n \to \infty$, $\frac{N}{n} \to 0$ bzw. $\frac{N_c}{n} \to 0$ aus den exakten Lösungen erschlossen werden. Zu diesem Zweck betrachten wir ein eindimensionales Modell, eine lineare Kette aus n Elementen, in dem keine schädigungsbedingten Spannungsumverteilungen auftreten: In allen (noch nicht gebrochenen) Elementen herrscht die gleiche Spannung σ. Die Elementeausfälle werden folglich durch eine Burchspannungsverteilung $F(\sigma)$ beschrieben, die für alle Elemente übereinstimmt. Zu einer Rißgruppe werden alle möglichen Risse mit dem gleichen (linken) Randpunkt zusammengefaßt. Für die Rißwahrscheinlichkeiten erhält man sofort

$$p_0 = 1 - F + F^2, \qquad p_N = F^N(1 - F)^2, \qquad (N \geqq 1), \qquad Q_{N_c} = F^{N_c}(1 - F). \tag{4}$$

Die Wahrscheinlichkeit p_0 ergibt sich dabei aus der Forderung, daß das betreffende Element entweder nicht gebrochen ist oder, wenn es gebrochen ist, nicht das linke Randelement eines Risses darstellen darf, d. h. auch das links angrenzende Element muß gebrochen sein.

Der direkte Weg zur exakten Berechnung des Rißspektrums ν_N und der makroskopischen Bruchwahrscheinlichkeit $F_A(\sigma)$ für das gegebene eindimensionale System führt über die allgemeine Strukturwahrscheinlichkeit $p_n(q_1, q_2, \dots, q_n)$ — die Wahrscheinlichkeit dafür, eine Realisierung des eindimensionalen Systems mit q_1 1-Rissen, q_2 2-Rissen usw. anzutreffen.

Aus der Strukturwahrscheinlichkeit lassen sich Rißspektrum und makroskopische Bruchwahrscheinlichkeit über folgende Beziehungen gewinnen:

$$\overline{q_M(n)} = \sum q_M p_n(q_1, \dots, q_n) , \tag{5}$$

$$\overline{\Delta q_M(n)^2} = \sum \left(q_M(n) - \overline{q_M(n)}\right)^2 p_n(q_1, \dots, q_n) , \tag{6}$$

$$F_A = F_{N_c}(n) = 1 - \sum_{q_1=0}^{n} \dots \sum_{q_{N_c-1}=0}^{n} p_n(q_1, \dots, q_{N_c-1}, 0, \dots, 0) . \tag{7}$$

Für eine offene Kette aus n Elementen erhält man (Anhang 1)

$$p_n(q_1, \dots, q_n) = \frac{(n - \Sigma N q_N + 1)!}{\Pi q_N! \left(n - \Sigma(N+1) q_N + 1\right)!} F^{\Sigma N q_N}(1-F)^{n-\Sigma N q_N} . \tag{8}$$

Die geschlossene Kette liefert wegen der geänderten Anschlußbedingungen einen leicht modifizierten Ausdruck (Anhang 1):

$$p_n(q_1, \dots, q_n = 0) = n \frac{(n - \Sigma N q_N - 1)!}{\Pi q_N! \left(n - \Sigma(N+1) q_N\right)!} F^{\Sigma N q_N}(1-F)^{n-\Sigma N q_N} , \tag{9}$$

$$p_n(0, \dots, 0, q_n = 1) = F^n .$$

Beim Übergang von der Strukturwahrscheinlichkeit $p_n(q_1, \dots, q_n)$ zu den gesuchten Größen ν_N und F_A sind folglich zahlreiche, durch den jeweiligen kombinatorischen Faktor gekoppelte Summationen auszuführen. HUNT [29], der die Bruchwahrscheinlichkeit F_A für die geschlossene Kette behandelte, bleibt — von einer zu (9) äquivalenten Wahrscheinlichkeit ausgehend — deshalb auf die numerische Auswertung angewiesen. Der numerische Aufwand steigt jedoch mit der Elementezahl n stark an, so daß dieses Verfahren von vornherein die Beschränkung auf nicht zu große Systeme auferlegt. Außerdem ist der Grenzübergang zu den Beziehungen (1) bis (3) nicht analytisch nachvollziehbar. Diese Schwierigkeiten lassen sich vermeiden, wenn man sich von vornherein auf die erforderlichen reduzierten Wahrscheinlichkeiten beschränkt. In [30] werden für die zu verschiedenen Systemgrößen gehörenden Wahrscheinlichkeiten $W_M(q_M, n)$ bzw. $F_{N_c}(n)$ Rekursionsgleichungen aufgestellt, aus denen sich mit Hilfe von Erzeugenden Funktionen die gewünschten Größen ableiten lassen.

Die genauere Betrachtung lehrt jedoch, daß die direkte Ausführung der gekoppelten Summationen in (5) bis (7) mit den Strukturwahrscheinlichkeiten (8) bzw. (9) durchaus möglich ist. Den Schlüssel hierfür liefert die im Anhang 2 abgeleitete Beziehung

$$\sum_{\substack{m_1=0 \\ \left(\sum_{j=1}^{c} j m_j = a\right)}}^{\infty} \dots \sum_{m_c=0}^{\infty} \frac{1}{\prod_{j=1}^{c} m_j!} \frac{1}{\left(b - \sum_{j=1}^{c} m_j\right)!} \Theta = \frac{1}{(b-1)!} \sum_{l=0}^{\infty} \frac{(a + b - (c+1) l - 1)!}{(a - (c+1) l)! \, l! \, (b-l)!} \Theta (-1)^l = \delta_{a0} \tag{10}$$

$$(b \geqq 1, c \geqq 1) \qquad (b = 0, c \geqq 1) .$$

Hierbei bezeichnet Θ symbolisch die Einschränkung der Summationen auf solche Werte der laufenden Variablen, für die die Argumente der Fakultäten nicht negativ sind. Durch (10) wird eine Vielzahl von Summationen, die über die Nebenbedingungen miteinander verkoppelt sind, auf eine einzige Summe zurückgeführt. Für $a \leqq c$ reduzieren sie sich auf das erste Glied mit $l = 0$:

$$\sum_{\substack{m_1=0 \\ \left(\sum_{j=1}^{c} j m_j = a\right)}}^{\infty} \dots \sum_{m_c=0}^{\infty} \frac{1}{\prod_{j=1}^{c} m_j!} \frac{1}{\left(b - \sum_{j=1}^{c} m_j\right)!} \Theta = \frac{(a+b-1)!}{a! \, b! \, (b-1)!} \Theta \qquad (a \leqq c) . \tag{11}$$

Mit Hilfe der Gleichungen (10) und (11) werden in den folgenden beiden Abschnitten Mittelwert und Schwankung der Rißzahl für die offene und geschlossene Elementekette berechnet. Anschließend wird das Verhalten dieser Größen in makroskopischen Systemen mit $n \gg 1$ und ihr Zusammenhang mit der Einzelrißnäherung behandelt. Den Abschluß bildet eine Diskussion der Abweichungen der genäherten Werte von denen der exakten Rechnung, die durch die Vernachlässigung der Randeffekte und der geometriebedingten statistischen Abhängigkeiten zwischen den Rissen hervorgerufen werden.

3. Offene Kette

Die mittlere Rißdichte $\overline{q_M}$ berechnet sich nach (5) und (8) aus

$$\overline{q_M(n)} = \sum_{q_1=0}^{n} \dots \sum_{q_n=0}^{n} F^{\Sigma N q_N} (1-F)^{n - \Sigma N q_N} \frac{(n + 1 - \Sigma N q_N)!}{\Pi q_N! (n + 1 - \Sigma N q_N - \Sigma q_N)!} q_M \Theta$$

$$= \sum_{B=0}^{\infty} F^B (1-F)^{n-B} (n + 1 - B)! \sum_{\substack{q_1=0 \\ (\Sigma N q_N = B)}}^{\infty} \dots \sum_{q_n=0}^{\infty} \frac{q_M}{\Pi q_N! (n + 1 - B - \Sigma q_N)!} \Theta . \tag{12}$$

Ersichtlich handelt es sich gerade um eine Summation des angegebenen Typs, wenn man m_j mit q_N für $N \neq M$ bzw. mit $q_M - 1$ identifiziert, sowie a mit $B - M$, b mit $n - B$ und c mit n. Wegen $\sum q_N N \leqq n$ ist die Ungleichung $a \leqq c$ stets erfüllt und es gilt

$$\overline{q_M(n)} = \sum_{B=0}^{\infty} F^B(1-F)^{n-B}\,(n+1-B)!\,\frac{(n-M-1)!}{(B-M)!\,(n-B)!\,(n-B-1)!}\,\Theta, \tag{13}$$

$$\overline{q_M(n)} = \sum_{B=M}^{n-1} F^B(1-F)^{n-B}\,(n+1-B)\binom{n-M-1}{B-M}, \qquad M \leqq n-1,$$

$$\overline{q_M(n)} = F^M(1-F)^{n-M} \sum_{k=0}^{n-M-1} (n-M+1-k)\binom{n-M-1}{k}\left(\frac{F}{1-F}\right)^k.$$

Durch wiederholte Ableitung der geometrischen Summe verifiziert man leicht die Beziehung

$$\sum_{k=0}^{l} (m-k)\binom{l}{k} x^k = (1+x)^{l-1}\,\{m(1+x) - lx\}. \tag{14}$$

Mit $x = F/(1-F)$, $l = n - M - 1$, $m = n - M + 1$ folgt

$$\overline{q_M(n)} = F^M(1-F)^{n-M}\left(1+\frac{F}{1-F}\right)^{n-M-2}\left\{(n-M+1)\left(1+\frac{F}{1-F}\right) - (n-M-1)\frac{F}{1-F}\right\}, \tag{15}$$

$$\overline{q_M(n)}\; F^M(1-F)\,\{(n-M+1) - (n-M-1)\,F\}, \qquad \overline{q_n(n)} = F^n.$$

Eine ganz analoge Rechnung ergibt für die mittlere quadratische Schwankung der Rißdichte (Anhang 3):

$$\begin{aligned}\overline{\Delta q_M(n)^2} &= (n-M+1)\,F^M - 2(n-M)\,F^{M+1} + (n-M-1)\,F^{M+2} - \big(n(2M+1) - 3M^2 + 1\big)\,F^{2M} + \\ &+ \big(2n(4M+3) - 4M(3M+2)\big)\,F^{2M+1} - \big(12n(M+1) - 6M(3M+4) - 8\big)\,F^{2M+2} + \\ &+ \big(2n(4M+5) - 12M(M+2) - 12\big)\,F^{2M+3} - \big(n(2M+3) - M(3M+8) - 5\big)\,F^{2M+4}.\end{aligned} \tag{16}$$

Zur Berechnung der Bruchwahrscheinlichkeit F_{N_c} ist nach (7) und (8) die Summe

$$\begin{aligned}1 - F_{N_c}(n) &= \sum_{q_1=0}^{n}{}' \cdots \sum_{q_{N_c-1}=0}^{n'} p_n(q_1, \dots, q_{N_c-1}, 0, \dots, 0) \\ &= \sum_{B=0}^{\infty} F^B(1-F)^{n-B}\,(n+1-B)! \sum_{\substack{q_1=0\\(\Sigma N q_N = B)}}^{\infty} \cdots \sum_{q_{N_c-1}=0}^{\infty} \frac{1}{\prod\limits_{N=1}^{M-1} q_N!\left(n+1-B-\sum\limits_{N=0}^{N_c-1} q_N\right)}\,\Theta;\end{aligned} \tag{17}$$

auszuwerten.

Die Anwendung des Hilfssatzes (10) erlaubt die weitgehende Ausführung der inneren Summationen:

$$\begin{aligned}1 - F_{N_c}(n) &= \sum_{B=0}^{\infty} F^B(1-F)^{n-B}\,\frac{(n+1-B)!}{(n-B)!} \sum_{l=0}^{\infty} \frac{(n-N_c l)!}{(B-N_c l)!\; l!\;(n+1-B-l)!}\,(-1)^l\,\Theta \\ &= \sum_{l=0}^{\infty} (1-F)^n\,(-1)^l\,\frac{(n-N_c l)!}{l!\,(n+1-(N_c+1)\,l)!}\,\Theta \sum_{B=N_c l}^{n+1-l} (n+1-B)\binom{n+1-(N_c+1)\,l}{B-N_c l}\left(\frac{F}{1-F}\right)^B,\end{aligned} \tag{18}$$

$$\begin{aligned}1 - F_{N_c}(n) &= \sum_{l=0}^{\infty} (1-F)^n\left(\frac{F}{1-F}\right)^{N_c l}(-1)^l\,\frac{(n-N_c l)!}{l!\,(n+1-(N_c+1)\,l)!}\cdot\Theta\,\times \\ &\times \sum_{A=0}^{n+1-(N_c+1)l} (n+1-N_c l - A)\binom{n+1-(N_c+1)\,l}{A}\left(\frac{F}{1-F}\right)^A.\end{aligned} \tag{19}$$

Die letzte Summe läßt sich mittels der Hilfsformel (14) ausführen:

$$\begin{aligned}1 - F_{N_c}(n) &= \sum_{l=0}^{\infty} (1-F)^n\left(\frac{F}{1-F}\right)^{N_c l}(-1)^l\,\frac{(n-N_c l)!}{l!\,(n+1-(N_c+1)\,l)!}\,\Theta\,\frac{1}{(1-F)^{n-(N_c+1)\,l+1}}\,\times \\ &\times \{(n+1-N_c l) - (n+1-(N_c+1)\,l)\,F\}.\end{aligned} \tag{20}$$

$$1 - F_{N_c}(n) = \sum_{l=0}^{\infty} (-1)^l F^{N_c l}(1-F)^{l-1}\left\{\binom{n+1-N_c l}{l} - F\binom{n-N_c l}{l}\right\}$$

oder

$$F_{N_c}(n) = -\frac{1}{1-F}\sum_{l=0}^{\infty}\big(-(F^{N_c}(1-F))\big)^l\left\{\binom{n+1-N_c l}{l} - F\binom{n-N_c l}{l}\right\}. \tag{21}$$

Die aus der Strukturwahrscheinlichkeit $p_n(q_1, q_2, \dots)$ so durch direkte Absummation errechneten Wahrscheinlichkeiten und Mittelwerte stimmen mit den aus speziellen Rekursionsgleichungen abgeleiteten Größen überein [30].

4. Geschlossene Kette

Die Berechnung der mittleren Rißdichte $\bar{q}_M$ und der Bruchwahrscheinlichkeit F_{N_c} für die geschlossene Kette verläuft ganz analog zu dem im vorangegangenen Abschnitt behandelten Fall. Nach Einsetzen der Strukturwahrscheinlichkeit (9) in (5) erhält man (für $M \neq n$)

$$q_M(n) = n(1-F)^n \sum_{B=0}^{n-1}\left(\frac{F}{1-F}\right)^B (n-B-1)! \sum_{\substack{q_1=0\\(\Sigma N q_N = B)}}^{\infty} \cdots \sum_{q_n=0}^{\infty} \frac{q_M\,\Theta}{\prod q_N!\,(n-B-\Sigma q_N)!}. \tag{22}$$

Die Anwendung von (11) auf die Vielfachsumme liefert

$$\overline{q_M(n)} = n(1-F)^n \sum_{B=N}^{n-2} \left(\frac{F}{1-F}\right)^B \frac{(n-N-2)!}{(n-B-2)!\,(B-N)!}\,, \qquad M < n-1\,, \tag{23}$$
$$\overline{q_{n-1}(n)} = nF^{n-1}(1-F)\,.$$

Nach einigen Vereinfachungen unter Ausnutzung der Beziehung (14) ergibt sich

$$\overline{q_M(n)} = nF^M(1-F)^2\,, \qquad M < n-1\,, \qquad \overline{q_{n-1}(n)} = nF^{n-1}(1-F)\,, \qquad \overline{q_n(n)} = F^n\,. \tag{24}$$

Für die mittlere quadratische Schwankung $\overline{\Delta q_M{}^2}$ erhält man nach den in Anhang 3 skizzierten Rechnungen

$$\overline{\Delta q_M(n)^2} = n\{F^M - 2F^{M+1} + F^{M+2} - (2M+1)\,F^{2M} + (8M+6)\,F^{2M+1} - 12(M+1)\,F^{2M+2} + (8M+10)\,F^{2M+3} - (2M+3)\,F^{2M+4}\}\,. \tag{25}$$

Zur Berechnung der Bruchwahrscheinlichkeit gehen wir von Gleichung (7) aus:

$$1 - F_{N_c}(n) = n(1-F)^n \sum_{B=0}^{\infty} \left(\frac{F}{1-F}\right)^B (n-B-1)! \sum_{\substack{q_1=0 \\ (\Sigma N q_N = B)}}^{\infty} \cdots \sum_{q_{N_c-1}=0}^{\infty} \frac{\Theta}{\Pi q_N!\,(n-B-\Sigma q_N)!}\,. \tag{26}$$

Die Vielfachsumme wird durch Anwendung der Beziehung (10) in eine Einfachsumme überführt:

$$1 - F_{N_c}(n) = n(1-F)^n \sum_{B=0}^{\infty} \left(\frac{F}{1-F}\right)^B \frac{(n-B-1)!}{(n-B-1)!} \sum_{l=0}^{\infty} \Theta \frac{(-1)^l\,(n-N_cl-1)!}{(B-N_cl)!\,l!\,(n-B-l)!}\,,$$
$$1 - F_{N_c}(n) = n(1-F)^n \sum_{B=0}^{\infty} \left(\frac{F}{1-F}\right)^B \sum_{l=0}^{\infty} \Theta \frac{(-1)^l\,(n-N_cl-1)!}{(B-N_cl)!\,l!\,(n-B-l)!} - F^n\,. \tag{27}$$

Beim letzten Schritt war zu beachten, daß durch das Weglassen der Fakultäten $(n-B-1)!$ zugleich die in Θ enthaltene Nebenbedingung $B \leq n-1$ entfiel und dadurch die Summation bis $B = n$ erstreckt wird. Der entsprechende Summand liefert gerade F^n und muß wieder abgezogen werden.

Durch Vertauschung der Reihenfolge der Summationen läßt sich eine der verbleibenden Summen auswerten:

$$1 - F_{N_c} = n(1-F)^n \sum_{l=0}^{\infty} \Theta \frac{(-1)^l\,(n-N_cl-1)!}{l!\,(n-(N_c+1)\,l)!} \sum_{B=N_cl}^{n} \binom{n-(N_c+1)\,l}{B-N_cl} \left(\frac{F}{1-F}\right)^B - F^n\,,$$
$$1 - F_{N_c} = n(1-F)^n \sum_{l=0}^{\infty} \Theta \frac{(-1)^l\,(n-N_cl-1)!}{l!\,(n-(N_c+1)\,l)!} \left(\frac{F}{1-F}\right)^{N_cl} \left(1+\frac{F}{1-F}\right)^{n-(N_c+1)l} - F^n\,. \tag{28}$$

Beachtet man noch, daß der Summand mit $l = 0$ gerade 1 liefert, so ergibt sich schließlich

$$F_{N_c} = -\sum_{l=0}^{\infty} \binom{n-N_cl}{l} \frac{n}{n-N_cl} \left(-F^{N_c}(1-F)\right)^l + F^n\,. \tag{29}$$

Abb. 1 zeigt den Verlauf der Bruchwahrscheinlichkeiten F_{N_c} in Abhängigkeit von der Elementebruchwahrscheinlichkeit F für ein kleines System aus $n = 10$ Elementen. Er stimmt bei der geschlossenen Kette mit den von HUNT [29] für diesen Spezialfall angegebenen numerischen Ergebnissen überein.

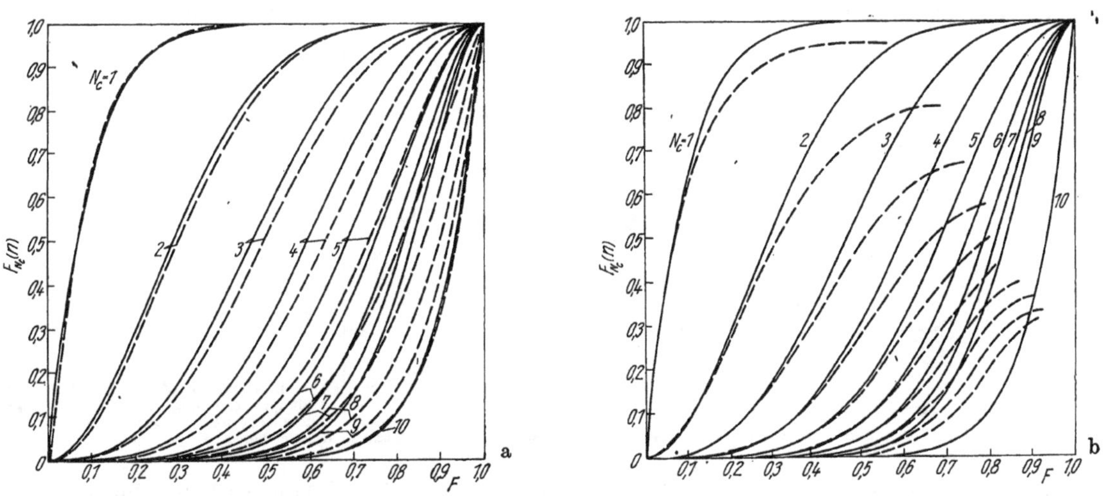

Abb. 1 Makroskopische Bruchwahrscheinlichkeit $F_A = F_{N_c}$ in Abhängigkeit von der Elementebruchwahrscheinlichkeit F für verschiedene kritische Rißgrößen N_c ($n = 10$).
a) geschlossene Kette (———) und offene Kette (– – –); b) geschlossene Kette, exakt (———) und Einzelrißnäherung (– – –)

5. Asymptotische Gültigkeit der Einzelrißnäherung

Es soll jetzt das Verhalten der Mittelwerte und Wahrscheinlichkeiten (15), (16) und (21) für die offene bzw. (24), (25) und (29) für die geschlossene Kette im Grenzfall großer Systeme mit $n \gg 1$, $n \gg N_c$ untersucht und mit den der Einzelrißnäherung entsprechenden Ausdrücken (1) bis (4) verglichen werden.

Die mittlere Rißzahl $\overline{q_M(n)}$ nimmt nach (15) bei der offenen Elementekette asymptotisch den mit der Näherung übereinstimmenden Wert

$$\overline{q_M(n)} \approx n(1 - F)^2 F^M \equiv q_{as,\,M}(n) \tag{30}$$

an. In Abb. 2 ist das Verhältnis der mittleren Rißzahl (15) zur asymptotischen Näherung $\overline{q_{as}}$ dargestellt. Bei kleinen Systemen treten zwar große Abweichungen auf, die jedoch mit wachsendem n schnell kleiner werden. Für $n \gtrsim 20M$ bleibt der Unterschied kleiner als 10%. Für die geschlossene Kette stimmt die Näherung von vornherein mit dem exakten Ausdruck (24) überein. Die Gleichsetzung der relativen Rißhäufigkeit $\nu_M \approx 1/n\,\overline{q_M(n)}$ mit der Rißwahrscheinlichkeit p_M wird also für große Systeme voll bestätigt.

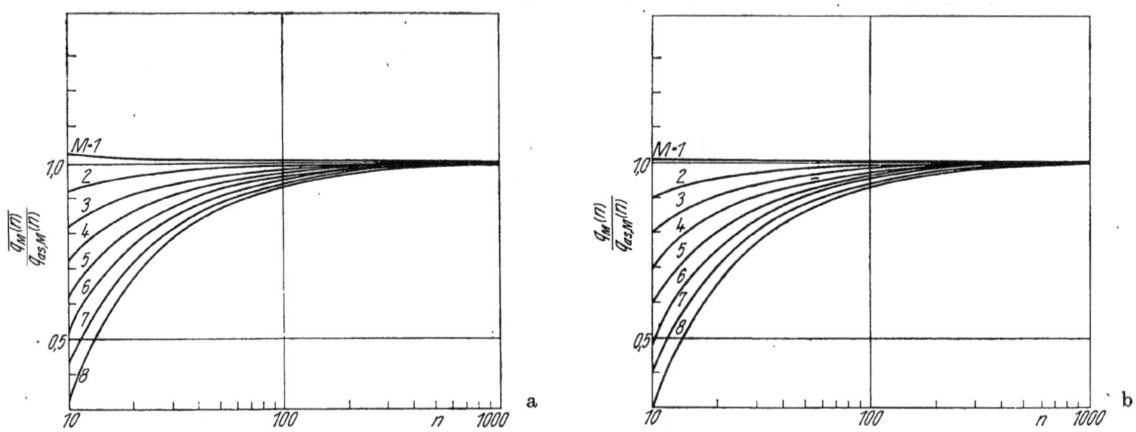

Abb. 2. Vergleich der mittleren Rißhäufigkeit $q_M(n)$ für die offene Elementekette nach (15) mit der asymptotischen Näherung $q_{as,M}(n) = nF^M(1 - F)^2$ nach (1) und (4).
a) $F = 0{,}01$; b) $F = 0{,}1$

Die quadratische Schwankung der Rißzahl einer offenen Elementekette wird asymptotisch proportional zur Elementezahl:

$$\Delta q_M(n)^2 = nF^M(1 - F)^2 + nF^{2M}\{-(2M - 1) + (8M + 6)\,F - (12M + 12)\,F^2 + (8M + 10)\,F^3 - (2M + 3)\,F^4\}\,. \tag{31}$$

Bei der geschlossenen Kette gilt diese Proportionalität nach (25) für beliebige Systemgrößen. Die relative Schwankung

$$\sqrt{\overline{\Delta q_M(n)^2}}\,/\overline{q_M(n)} \sim 1/\sqrt{nF^M} \qquad (\text{für } n \gg M) \tag{32}$$

strebt folglich für große Systeme gegen Null. Die Zahl $q_M(n)$ der M-Risse kann also in makroskopischen Systemen — wie es in (1) behauptet wurde — in guter Näherung durch die mittlere Rißzahl $\overline{q_M(n)}$ ersetzt werden. (Abb. 3). Die Systeme müssen allerdings so groß sein, daß sich zahlreiche M-Risse bilden und damit die Schwankungen weitgehend herausmitteln können. Dazu muß die Bedingung $\overline{q_M(n)} \approx nF^M \gg 1$ erfüllt sein.

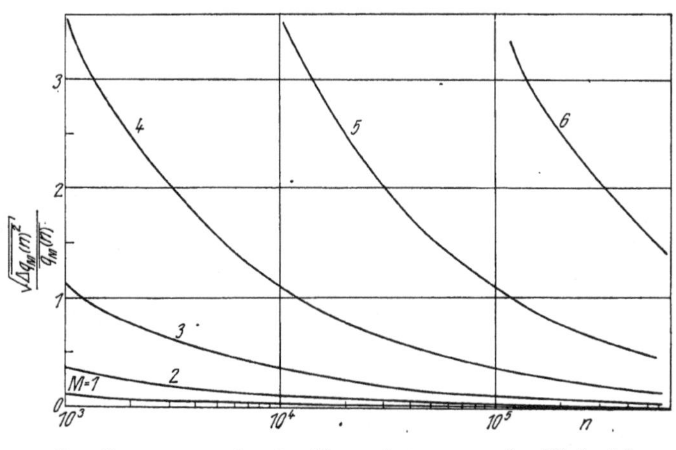

Abb. 3. Abhängigkeit der relativen Schwankung der Rißlängenverteilung von der Systemgröße n ($F = 0{,}1$) (Die offene und die geschlossene Kette liefern praktisch die gleichen Kurven.)

In sehr großen Systemen mit $n \gg N_c$ genügt wegen der Vielzahl an möglichen Rißbildungen schon eine kleine Elementebruchwahrscheinlichkeit F, um den Gesamtbruch mit einer merklichen Wahrscheinlichkeit $F_A = F_{N_c}(n)$ auszulösen. Infolge des Vorfaktors $F^{N_c l}$ wird deshalb der Hauptbeitrag von den Summanden mit $N_c\, l \ll n$ getragen.

Mit der hieraus folgenden Näherung

$$\binom{n - N_c l + 1}{l} \approx \binom{n}{l}, \qquad \binom{n - N_c l}{l} \approx \binom{n}{l} \tag{33}$$

gehen die exakten Ausdrücke (21) und (29) für die makroskopische Bruchwahrscheinlichkeit in

$$\begin{aligned} F_{N_c}(n) &\approx 1 - \sum_{l=0}^{\infty} \binom{n}{l} \left(-F^{N_c}(1 - F)\right)^l = 1 - \left(1 - (1 - F)\, F^{N_c}\right)^n , \\ F_{N_c}(n) &\approx 1 - (1 - Q_{N_c})^n \end{aligned} \tag{34}$$

über. Auch hier ergibt sich im Grenzfall einer großen Elementezahl das Resultat der Einzelrißnäherung, wobei — gemäß dem Kriterium $n \gg N_c$ — die Übereinstimmung für kleine N_c und kleine F immer besser wird.

6. Einfluß des Randes und der geometrischen Korrelationen

Bei der Einzelrißnäherung werden die statistischen Abhängigkeiten, die zwischen verschiedenen Rissen bestehen, vernachlässigt. Außerdem bleiben etwaige Randeffekte außer Betracht. Der Einfluß dieser beiden Erscheinungen soll im folgenden diskutiert werden.

Mit der geschlossenen Kette liegt ein System vor, in dem durch die periodischen Anschlußbedingungen die Sonderrolle der Randelemente beseitigt ist. Die durch den Rand bedingten Veränderungen $\delta_R A$ einer Größe A lassen sich folglich aus dem Vergleich der offenen und der geschlossenen Kette erschließen:

$$\delta_R A = A_{\text{offene Kette}} - A_{\text{geschlossene Kette}} .$$

Für Mittelwert und Schwankung der Rißzahl $q_M(n)$ erhält man aus (15), (16), (24) und (25):

$$\delta_R \overline{q_M(n)} = -F^M(1 - F) \left\{(M - 1) - (M + 1) F\right\} , \tag{35}$$

$$\begin{aligned} \delta_R \, \Delta q_M(n)^2 = &-F^M\{(M - 1) - 2MF + (M + 1) F^2\} + \\ &+ F^{2M}\{3M^2 - 1 - 4M(3M + 2) F + (18M^2 + 24M + 8) F^2 - 12(M + 1)^2 F^3 + (3M^2 + 8M + 5)F^4\} . \end{aligned} \tag{36}$$

Diese Differenzen hängen nicht von der Gesamtzahl n der Elemente ab. Es ist plausibel, daß Veränderungen der Rißzahl, die durch Modifikationen an der Systembegrenzung bedingt sind, nicht durch die Vergrößerung des Systems beeinflußt werden. Da Mittelwerte und Schwankungen proportional zur Systemgröße anwachsen, verschwinden die relativen, randbedingten Unterschiede im entsprechenden Maße. Durch den Rand wird bei der offenen Kette die weitere Ausbreitung randnaher Risse verhindert. Durch diesen Randeinfluß wird deshalb die mittlere Rißzahl und (nach (36)) auch die Schwankung vermindert. Dies kommt (wegen $q_{as} = \overline{q}_{\text{geschlossene Kette}}$) unmittelbar in Abb. 2 zum Ausdruck.

Die statistischen Korrelationen zwischen den Rissen werden durch die Überlappung der verschiedenen möglichen Rißbildungen verursacht. Ein und dasselbe gebrochene Element kann zu unterschiedlichen Rissen gehören und so deren Wahrscheinlichkeiten miteinander verkoppeln. Zur Abschätzung dieses Effektes konstruieren wir die Wahrscheinlichkeiten für den Fall, daß die Bildung der verschiedenen Risse einer bestimmten Größe M als statistisch unabhängige Ereignisse angesehen werden können. Der Einfluß der geometrisch bedingten Korrelationen auf eine Größe A kann dann aus der Differenz $\delta_K A = A(\text{exakt}) - A$ (statistisch unabhängige Risse) abgeschätzt werden. Im Falle der offenen Kette erhält man bei vorausgesetzter statistischer Unabhängigkeit für die Wahrscheinlichkeit $W_M(q_M, n)$, daß im System aus n Elementen q_M Risse der Größe M existieren:

$$\begin{aligned} W_M(q_M, n) = &\binom{n - M - 1}{q_M} P_M^{q_M} (1 - P_M)^{n-M-1-q_M} (1 - P'_M)^2 + \\ &+ 2\binom{n - M - 1}{q_M - 1} P_M^{q_M - 1} (1 - P_M)^{n-M-q_M} P'_M (1 - P'_M) + \\ &+ \binom{n - M - 1}{q_M - 2} P_M^{q_M - 2} (1 - P_M)^{n-M-q_M+1} P'^2_M . \end{aligned} \tag{37}$$

Dabei sind unter P_M die Wahrscheinlichkeiten für die Bildung eines M-Risses im Innern der Kette und unter P'_M die Wahrscheinlichkeiten der Randrisse zu verstehen:

$$P_M = F^M(1 - F)^2 , \qquad P'_M = F^M(1 - F) . \tag{38}$$

Der erste Summand in (37) erfaßt die Fälle, bei denen alle M-Risse im Innern liegen, beim zweiten und dritten Summanden liegen ein bzw. zwei M-Risse an den Begrenzungen. Bei der geschlossenen Elementekette braucht diese Unterscheidung nicht getroffen zu werden und man erhält sofort die Binomialverteilung

$$W_M(q_M, n) = \binom{n}{q_M} P_M^{q_M}(1 - P_M)^{n-q_M} . \tag{39}$$

Die Berechnung der mittleren Rißzahl nach

$$\overline{q_M(n)} = \sum_{q_M} q_M W_M(q_M, n) \tag{40}$$

liefert

$$\overline{q_M(n)} = (n - M - 1) P_M + 2P'_M \qquad \text{(offene Kette)},$$
$$\overline{q_M(n)} = nP_M \qquad \text{(geschlossene Kette)}. \tag{41}$$

Bei Beachtung von (38) stimmen überraschenderweise die Mittelwerte vollständig mit den exakten Ausdrücken (15) und (24) überein. Die statistischen Korrelationen zwischen den Rissen haben also auf deren mittlere Häufigkeit keinen Einfluß. Offensichtlich kompensieren sich ihre beiden entgegengerichteten Wirkungen genau: Die Tatsache, daß ein gebrochenes Element nur zu einem einzigen realisierten M-Riß gehören darf, verringert ihre Gesamtzahl. Zugleich aber kann ein nicht gebrochenes Element gleichzeitig zwei Risse begrenzen. Die statistischen Abhängigkeiten zwischen den Rissen machen sich erst im zweiten Moment der $W_M(q_M, n)$-Verteilung und damit in der mittleren quadratischen Schwankung $\overline{\Delta q_M(n)^2}$ bemerkbar. Für die offene Kette ergibt sich aus (37) nach einer längeren, aber elementaren Rechnung

$$\overline{\Delta q_M(n)^2} = (n - M - 1) P_M(1 - P_M) + 2P'_M(1 - P'_M). \tag{42}$$

Durch Einsetzen von (38) und Vergleich mit (16) erhält man für den durch die geometrischen Korrelationen bewirkten Einfluß

$$\delta_K \overline{\Delta q_M(n)^2} = -F^{2M}\{n\,2M - 3M^2 + M - \big(n(8M + 2) - 12M^2 - 4M\big) F + \big(n(12M + 6) - 18M^2 - 18M - 4\big) F^2 - \big(n(8M + 6) - 12M^2 - 20M - 8\big) F^3 + \big(n(2M + 2) - 3M^2 - 7M - 4\big) F^4\}. \tag{43}$$

Im Fall der geschlossenen Kette erhält man aus (39) für die Schwankungen ohne Berücksichtigung der Korrelationen zwischen den Rissen

$$\overline{\Delta q_M(n)^2} = nP_M(1 - P_M). \tag{44}$$

Die Abweichung des exakten Resultates (25) beträgt

$$\delta_K \overline{\Delta q_M(n)^2} = -nF^{2M} \{2M - (8M + 2) F + (12M + 6) F^2 - (8M + 6) F^3 + (2M + 2) F^4\}$$
$$= -nF^{2M}(1 - F)^3\, 2\big(N(1 - F) - F\big). \tag{45}$$

Bei großen Systemen mit $n \gg M$ stimmen die Differenzen (43) und (45) für die offene und die geschlossene Kette miteinander überein: Die Korrelationen zwischen den Rissen liefern dann unabhängig von den speziellen Randbedingungen den gleichen Beitrag zu $\overline{\Delta q_M(n)^2}$. Er ist (wie $\overline{\Delta q_M(n)^2}$ selbst) im wesentlichen proportional zu F^{2M} und bildet damit — wie es bei einem das Ketteninnere betreffenden Effekt zu erwarten ist — einen von der Systemgröße unabhängigen relativen Anteil. Die statistischen Abhängigkeiten zwischen den Rissen vermindern nach (45) die Schwankungen der Rißzahlen, so daß die in der Einzelrißnäherung gemachte Annahme $q_M(n) \approx \overline{q_M(n)}$ erst recht erfüllt ist (In größeren Systemen interessieren nur kleine Elementebruchwahrscheinlichkeiten F, bei denen die Bedingung $F < M/(M + 1)$ immer erfüllt ist).

7. Schlußfolgerungen

In der Einzelrißnäherung wird bei der statistischen Rißbildung- und -ausbreitung lediglich die Umgebung des einzelnen Risses in Betracht gezogen. Randeffekte und statistische Korrelationen zwischen benachbarten Rissen werden dabei vernachlässigt. Die in den vorangegangenen Abschnitten erhaltenen Ergebnisse für das Bruchverhalten in linearen, wechselwirkungsfreien Elementeketten lassen die folgenden Aussagen über diese Näherungen zu:

1. Die Einzelrißnäherung ist für die Berechnung der mittleren Rißhäufigkeit und der makroskopischen Bruchwahrscheinlichkeit anwendbar, wenn das System sehr viel größer als die maximale, kritische Rißgröße ist ($n \gg N_c$).
2. Die Schwankungen der Rißzahl können vernachlässigt werden, falls genügend Risse der betrachteten Sorte vorhanden sind ($nF^M \gg 1$).
3. Der Einfluß des Randes auf die Rißzahl ist unabhängig von der Systemgröße. Für Zweier- und größere Risse wird durch die Begrenzung die mittlere Rißzahl und die Schwankung vermindert.
4. Die statistischen Korrelationen beeinflussen die mittlere Rißzahl nicht (Da bei der geschlossenen Kette ohnehin keine Randeffekte zu berücksichtigen sind, liefert für diesen Fall die Einzelrißnäherung das exakte Resultat). Sie verringern die quadratische Schwankung in Proportionalität zur Systemgröße.

Diese Folgerungen wurden hier für den Fall des linearen Modellsystems bewiesen, doch sollten sie in der Tendenz auch für kompliziertere Systeme gelten.

Anhang 1: Berechnung der allgemeinen Strukturwahrscheinlichkeit $p_n(q_1, \dots, q_n)$ (Gleichung (8) und (9))

Die Realisierungswahrscheinlichkeit für eine konkrete Anordnung von $n' = \sum_N N q_N$ gebrochenen Elementen ist durch $F^{n'}(1-F)^{n-n'}$ gegeben. Die Strukturwahrscheinlichkeit $p_n(q_1, \dots, q_n)$ ergibt sich hieraus durch Multiplikation mit der Anzahl $C_n(q_1, \dots, q_n)$ der verschiedenen Möglichkeiten, q_1 1-Risse, q_2 2-Risse usw. im System aus n Elementen anzuordnen:

$$p_n(q_1, \dots, q_n) = C_n(q_1, \dots, q_n)\, F^{n'}(1-F)^{n-n'}\,.$$

Bei der offenen Kette wählen wir als unabhängig zu permutierende Komplexe die $k = \sum_N q_N$ Risse mit ihren rechten, nichtgebrochenen Nachbarelementen, sowie die restlichen $n + 1 - n' - k$ nichtgebrochenen Elemente (Das System muß dazu durch ein nichtgebrochenes Element am rechten Systemrand erweitert werden). Diese Komplexe können auf $(n + 1 - n')!$ verschiedene Weisen angeordnet werden. Da die q_N N-Risse und die $n + 1 - n' - k$ „freien" ungebrochenen Elemente nichtunterscheidbar sind, ist noch durch $\prod_N q_N!\,(n + 1 - n' - k)!$ zu dividieren:

$$C_n(q_1, \dots, q_n) = \frac{(n+1-n')!}{\prod_N q_N!\,(n+1-n'-k)!} = \frac{k!}{\prod_N q_N!}\binom{n+1-n'}{k}.$$

Diese Überlegungen gelten selbstverständlich nur für $n + 1 - n' \geqq 0$, andernfalls ist $C_n = 0$.

Bei der geschlossenen Kette verläuft die Berechnung des kombinatorischen Faktors ganz analog: Wir fassen wieder alle Risse einschließlich ihres rechten nichtgebrochenen Nachbarelementes zu einem Rißkomplex zusammen (Es wird $q_n = 0$ vorausgesetzt). Es liegen $k = \sum q_N$ derartige Rißkomplexe und $n - n' - k = n - \sum q_N\,(N + 1)$ nichtgebrochene, freie Elemente, also insgesamt $n - n'$ unabhängig zu permutierende Komplexe vor. Unter Beachtung der Nichtunterscheidbarkeit können sie auf $(n - n')!/\prod q_N! \times (n - n' - k)!$ verschiedene Arten angeordnet werden. Auf dem geschlossenen Ring sind dabei jeweils $(n - n')$ Permutationen, die sich nur durch eine zyklische Vertauschung unterscheiden, bezüglich der Reihenfolge der kombinatorischen Komplexe gleichwertig. Die Anzahl verringert sich folglich auf $(n - n' - 1)!/\prod q_N!\,(n - n' - k)!$. Jede solche spezielle Rißkonfiguration kann n verschiedene Lagen in der geschlossenen Kette annehmen, so daß wir schließlich

$$C_n(q_1, \dots, q_n) = n\,\frac{(n-n'-1)!}{q_N!\,(n-n'-k)!} = n\,\frac{(k-1)!}{\prod q_N!}\binom{n-n'-1}{k-1}$$

erhalten.

Anhang 2: Auswertung der gekoppelten Summationen (Gleichung (10))

Zur Auswertung der Summen

$$g_{ab} = \sum_{m_1=0}^{\infty} \cdots \sum_{m_c=0}^{\infty} \frac{1}{\prod\limits_{j=0}^{c} m_j!}\,\frac{1}{(b - \sum\limits_{j=1}^{c} m_j)!} \qquad (\Sigma m_j \leqq b)\,, \qquad (\Sigma m_j j = a)\,,$$

die über die Nebenbedingung miteinander gekoppelt sind, führen wir die erzeugende Funktion

$$\Phi(u, v) = \sum_{a=0}^{\infty}\sum_{b=0}^{\infty} u^a v^b g_{ab}$$

ein. Wegen der zusätzlichen Aufsummationen können die Nebenbedingungen fallen gelassen werden. Mit $m_0 = b - \sum\limits_{j=1}^{c} m_j$ erhält man

$$\Phi(u, v) = \sum_{m_0=0}^{\infty} \cdots \sum_{m_c=0}^{\infty} \frac{1}{\prod\limits_{j=0}^{c} m_j!}\, u^{\sum\limits_{j=0}^{c} m_j j}\, v^{\sum\limits_{j=0}^{c} m_j} = \prod_{j=0}^{c}\left(\sum_{m_j=0}^{\infty}\frac{(u^j v)^{m_j}}{m_j!}\right) = \prod_{j=0}^{c} e^{u^j v} = e^{v \cdot \frac{1-u^{c+1}}{1-u}}\,.$$

Die weitere Aufgabe besteht darin, $\Phi(u, v)$ in eine Potenzreihe nach u und v zu entwickeln. Ihre Koeffizienten stellen dann die gesuchten Ausdrücke g_{ab} dar. Es gilt

$$\Phi(u, v) = \sum_{b=0}^{\infty} v^b \frac{1}{b!}\left(\frac{1-u^{c+1}}{1-u}\right)^b.$$

Zur Auswertung der Potenzen benutzen wir ihre Darstellbarkeit als geometrische Reihe

$$\left(\frac{1-u^{c+1}}{1-u}\right)^b = \sum_{s=0}^{\infty}\binom{-b}{s}(-1)^s u^s \sum_{l=0}^{\infty}\binom{b}{l}(-1)^l u^{(c+1)l} = \sum_{s=0}^{\infty}\sum_{l=0}^{\infty} u^{s+(c+1)l}(-1)^l \frac{(b+s-1)!}{s!\,(b-1)!}\,\frac{b!}{l!\,(b-l)!}\,\Theta =$$

$$= b!\sum_{a=0}^{\infty} u^a \frac{1}{(b-1)!}\sum_{l=0}^{\infty}(-1)^l \frac{(a+b-(c+1)\,l-1)!}{(a-(c+1)\,l)!\,l!\,(b-l)!}\,\Theta\,.$$

Durch Θ wird der Abbruch der Summation bei negativen Argumenten der Fakultäten ausgedrückt. Einsetzen dieser Entwicklung in den obigen Ausdruck für die erzeugende Funktion und Koeffizientenvergleich mit ihrer Definitionsgleichung liefert die gesuchte Beziehung (10) für $b \geqq 1$ (Die benutzte Umformung des Binominalkoeffizienten $\binom{-b}{s}$ ist nur unter dieser Voraussetzung möglich).

Wir weisen auf die hieraus folgenden bemerkenswerten Identitäten

$$\sum_{l=0}^{\infty}(-1)^l \frac{(a+b-(c+1)\,l-1)!}{(a-(c+1)\,l)!\,l!\,(b-l)!}\,\Theta = 0 \qquad (a > bc,\ b \geqq 1,\ c \geqq 1)$$

hin. Für alle g_{ab} mit $a > b \cdot c$ (und $b, c \geqq 1$) läßt sich nämlich kein Wertesatz $(m_1, \dots, m_c)$ finden, der mit den Nebenbedingungen $\sum\limits_{j=1}^{c} m_j \leqq b$ und $\sum\limits_{j=1}^{c} m_j j = a$ verträglich ist.

Für $b = 0$ erhält man aus $\Phi(u, v)$ oder durch direkte Auswertung der Summen $g_{a0} = \delta_{a0}$.

Anhang 3: Berechnung der quadratischen Schwankung (Gleichung (16) und (25))

Die mittlere quadratische Schwankung $\overline{\Delta q^2}$ läßt sich in der Form

$$\overline{\Delta q^2} = \overline{(q - \bar{q})^2} = \overline{q(q-1)} - \bar{q}^2 + \bar{q}$$

darstellen. Bei der Berechnung des Mittelwertes

$$\overline{q_M(q_M - 1)} = \sum_{q_1} \dots \sum_{q_n} q_M(q_M - 1)\, p_n(q_1, \dots, q_n)$$

für die offene Kette gehen wir von den Wahrscheinlichkeiten (8) aus. Es gilt dann:

$$\overline{q_M(q_M - 1)} = (1-F)^n \sum_{B=0}^{\infty} \left(\frac{F}{1-F}\right)^B (n - B + 1)! \sum_{\substack{q_1=0 \\ (\Sigma N q_N = B)}}^{\infty} \dots \sum_{q_n=0}^{\infty} \frac{q_M(q_M-1)}{\Pi q_N!(n - B + 1 - \Sigma q_N)!}.$$

Bei der Anwendung von Gleichung (11) ist hier m_j mit q_N (für $N \neq M$) bzw. $q_M - 2$, b mit $n - B - 1$ und a mit $B - 2M$ zu identifizieren:

$$\overline{q_M(q_M - 1)} = (1-F)^n \sum_{B=0}^{\infty} \left(\frac{F}{1-F}\right)^B (n - B + 1)! \frac{(n - 2M - 2)!}{(n - B - 1)!\,(n - B - 2)!\,(B - 2M)!}$$

$$= (1-F)^n \sum_{B=0}^{\infty} \left(\frac{F}{1-F}\right)^B (n - B + 1)(n - B) \binom{n - 2M - 2}{B - 2M}.$$

Zur Auswertung benutzen wir die Beziehung

$$\sum_{k=0}^{n} (m + 1 - k)(m - k) \binom{n}{k} x^k = x^{m+2} \frac{d^2}{dx^2} \frac{(1+x)^n}{x^m} = (1+x)^{n-2} \{n(n-1)x^2 - mn2x(1+x) + m(m+1)(1+x)^2\}.$$

Sie gestattet die Umformung

$$\overline{q_M(q_M - 1)} = (1-F)^n \left(\frac{F}{1-F}\right)^{2M} \sum_{k=0}^{n-2M-2} (n - 2M + 1 - k)(n - 2M - k) \binom{n - 2M - 2}{k} \left(\frac{F}{1-F}\right)^k$$

$$= (1-F)^2 F^{2M} \{(n - 2M + 1)(n - 2M) - 2(n - 2M - 2)(n - 2M) F + (n - 2M - 2)(n - 2M - 3) F^2\}.$$

Zusammen mit den aus (15) zu entnehmenden Ausdrücken für $\bar{q}_M$ und $\bar{q}_M^2$ erhält man nach Zusammenfassung gleicher Potenzen in F die mittlere quadratische Schwankung gemäß (16).

Ganz analog verläuft die Rechnung im Falle der geschlossenen Kette mit der Wahrscheinlichkeit $p_n(q_1, \dots, q_n)$ nach Gleichung (9):

$$\overline{q_M(q_M - 1)} = n(1-F)^n \sum_{B=0}^{\infty} \left(\frac{F}{1-F}\right)^B (n - 1 - B)! \sum_{q_1=0}^{\infty} \dots \sum_{q_n=0}^{\infty} \frac{q_M(q_M-1)}{\Pi q_k!(n - B - \Sigma q_k)!}\, \Theta$$

$$= n(1-F)^n \sum_{B=0}^{\infty} \left(\frac{F}{1-F}\right)^B (n - 1 - B)! \frac{(n - 2M - 3)!}{(n - B - 2)!\,(n - B - 3)!\,(B - 2M)!}\, \Theta$$

$$= n(1-F)^n \left(\frac{F}{1-F}\right)^{2M} \sum_{k=0}^{n-2M-3} \left(\frac{F}{1-F}\right)^k (n - 1 - (k + 2M)) \binom{n - 2M - 3}{k}$$

$$= n(1-F)^3 F^{2M} \{(n - 2M - 1) - (n - 2M - 3) F\}.$$

Unter Berücksichtigung der Mittelwerte (24) folgt daraus der in (25) angegebene Ausdruck für die quadratische Schwankung.

Literatur

1 Peirce, F. T., Tensile tests for cotton yarns V., J. Textile Institute Transactions, **17** (1926) 355.
2a Weibull, W., A statistical theory of the strength of materials, Ingeniör Vetenskaps Akademien Handlingar, Nr. 151 (1939).
2b Weibull, W., The phenomenon of rupture in solids, Ingeniör Vetenskaps Akademien Handlingar, Nr. 153 (1939).
3 Kontorova, T. A., Frenkel, Ja. I., Statističeskaja teorija chrupkoj pročnosti realnych kristallov, Žurnal techniceskoj fiziki, **11** (1941) 173.
4a Bolotin, V. V., Statistische Methoden in der Baumechanik, Schriftenreihe der Bauforschung, Reihe Technik und Organisation **44**, Berlin 1972.
4b Bolotin, V. V., Primenenie metodov teorii verojatnostej i teorii nadežnostej v rasčetach sooruženij, Moskau 1971.
5 Schultrich, B., Gille, G., Pompe, W., Bruchstatistik I, Modell des schwächsten Kettengliedes, preprint TU/ZFW, Dresden 1976.
6 Daniels, H. E., The statistical theory of the strength of bundles of threads I, Proc. Royal Society A **183** (1945) 405.
7 Rosen, B. W., Tensile failure of fibrous composites, AIAA Journal, **2** (1964) 1985.
8 Phoenix, S. L., Probabilistic strength analysis of fibre bundle structures, Fibre Science and Technology, 7 (1974) 15.
9 Schultrich, B., Bruchstochastik, Dissertation B, TU Dresden 1977.
10a Bogačev, I. N., Vajnštejn, A. A., Volkov, S. D., Vvedenie v statističeskoe metallovedenie, Moskau 1972.
10b Volkov, S. D., Statistical strength theory, New York 1962.
11 Zhurkov, S. N., Kinetic concept of the strength of solids, Intern. J. Fracture Mechanics, **1** (1965) 311.
12 Regel, V. P., Slucker, A. I., Tomaševski, E. E., Kinetičeskaja priroda pročnosti tverdych tel, Moskau 1974.
13 Tamuž, V. P., Kuksenko, V. S., Mikromechanika razrušenija polimernych materialov, Riga 1978.
14 Hsieh, C., Thomson, R., Lattice theory of fracture and crack creep, J. Applied Physics, **44** (1973) 2051.
15 Lawn, B. R., An atomistic model of kinetic crack growth in brittle solids, J. Materials Science, **10** (1975) 469.
16a Petrov, V. A., Vladimirov, V. I., Orlov, A. N., A kinetic approach to fracture of solids, phys. stat. sol., **42** (1970) 197.
16b Orlov, A. N., Petrov, V. A., Vladimirov, V. I., A kinetic approach to fracture of solids II, phys. stat. sol. (b) **47** (1971) 293.
17 Petrov, V. A., Javlenija termofluktuacionnovo razrušenija, Fizika tverdovo tela, **18** (1976) 1290.
18 Charles, R. J., Static fatigue of glass II, J. Applied Physics, **29** (1958) 1554.
19 Evans, A. G., Wiederhorn, S. M., Proof testing of ceramic materials — an analytical basis for failure prediction, Intern. J. Fracture, **10** (1974) 379.
20 Posselt, M., Schultrich, B., Kinetische Theorie des zeitabhängigen Bruchverhaltens heterogener Materialien mit spröden Komponenten, TU-preprint 05-29-79, Dresden 1980.
21 Schultrich, B., Pompe, W., Bruchstatistik III: Grundlagen der Bruchstatistik für wechselwirkende Systeme, Teil 1, TU-preprint 05-34-78, Dresden 1978.

22a ZWEBEN, C., The strength of notched and damaged composites, J. Composite Materials, **3** (1969) 713.
22b ZWEBEN, C., ROSEN, B. W., A statistical theory of material strength with application to composite materials, J. Mech. Phys. Solids, **18** (1970) 189.
23 ARGON, A. S., Statistical Aspects of fracture, in BROUTMAN, L. J., KROCK, R. H. (Ed.), Composite Materials, Vol. V Fracture and Fatigue, New York 1974, S. 153.
24a SCHULTRICH, B., POMPE, W., Bruchstatistik III: Grundlagen der Bruchstatistik für wechselwirkende Systeme, Teil 2, TU-preprint 05-35-78, Dresden 1978.
24b SCHULTRICH, B., POSSELT, M., Bruchstatistik III: Grundlagen der Bruchstatistik für wechselwirkende Systeme, Teil 3, TU-preprint 05-22-79, Dresden 1980.
25 BUCH, J. D., ZIMMER, J. E., MEYER, R. A., An analytical microstructural model for the fracture of graphite, Report ATR-74 7425)-4, The Aerospace Corp., El Segundo (Cal.) 1974.
26 PALLEJ, I. Z., VAJC, G. A., K opredeleniju verojatnosti razrušenija reguljarnych kompozitionnych struktur, Voprosy dinamiki i pročnosti, **29** (1974) 138.
27 SCHULTRICH, B., GILLE, G., POMPE, W., Bruchstatistik II: Modelle ohne Feldwechselwirkung, preprint TU/ZFW, Dresden 1977.
28 ČUDNOVSKIJ, A. N., Voprosy uprugosti i plastičnosti, Leningrad 1973.
29 HUNT, R. A., A theory of the statistical linking of micro-cracks consistent with classical reliability theory, Acta Metallurgica, **26** (1978) 1443.
30 SCHULTRICH, B., POSSELT, M., Statistisches Bruchverhalten von linearen Strukturen, Wiss. Zeitschrift TU Dresden, im Erscheinen.

Eingereicht am 4. 7. 1979, in revidierter Fassung am 14. 4. 1980

Anschriften: Dr. sc. nat. BERND SCHULTRICH, Zentralinstitut für Festkörperphysik und Werkstofforschung, 8027 Dresden, Helmholtzstr. 20, DDR / Dipl.-phys. MATTHIAS POSSELT, Technische Universität Dresden, Sektion Physik, 8027 Dresden, Mommsenstr. 13, DDR

ZAMM 60, 569 – 575 (1980)

W. Kliem

Free and Forced Vibrations of a Linear Non-Conservative System with Multiple Eigenvalues

Die Lösungen des linearen Differentialgleichungssystems $A\ddot{q} + B\dot{q} + Cq = Q(t)$ werden untersucht. Hierbei seien A, B und C quadratische Matrizen mit reellen und konstanten Elementen, aber die Matrizen seien nicht notwendigerweise symmetrisch oder positiv definit. Zusätzlich wird angenommen, daß mehrfache Eigenwerte mit einer ungenügenden Anzahl von Eigenvektoren auftreten. Für einfache Eigenwerte ist eine entsprechende Theorie von Wahed und Bishop [1] aufgestellt worden.

The paper shows the derivation of the response of a linear system governed by the equation $A\ddot{q} + B\dot{q} + Cq = Q(t)$, in which the square matrices A, B and C are real and constant, but not necessary symmetric or positive definite. In addition the special case of multiple eigenvalues with an insufficient number of eigenvectors occurs. The theory is an extension of the work by Wahed and Bishop [1], which is limited to distinct eigenvalues.

Решение линейной системы дифференциальных уравнений $A\ddot{q} + B\dot{q} + Cp = Q(t)$ исследуется методом Лагранжа. A, B, и C — квадратные матрицы с действительными и постоянными элементами, но эти матрицы не обязательно симметричны или положительно определены. Дополнительно предполагается, что многократные собственные значения появляются с недостаточным количеством собственных векторов. Для простых собственных значений соответствующая теория развита Ваедом и Бишопом [1].

1. Introduction

In [1] Wahed and Bishop derive the solution of the equation

$$A\ddot{q} + B\dot{q} + Cq = Q(t)\,, \tag{1}$$

in which the mass matrix A, the damping matrix B and the stiffness matrix C are real, constant matrices of order $n \times n$ with non of the restrictions usual in modal analysis. $Q(t)$ is the excitation vector of order $n \times 1$ and q, likewise of order $n \times 1$, represents the generalized coordinates.

The strength of [1] results from the Lagrangian approach, i.e. the n differential equations of second order (1) are used directly, while other authors (e.g. [2, 3]) transform the system (1) into $2n$ equations of first order (Hamiltonian approach). In this way Wahed and Bishop preserve a series of well known characteristics of conventional linear vibration theory and produce a very application-minded method. Their method was used in the original form and variations of this by Fawzy and Bishop [4, 5], Bishop and Price [6] and Lund [7] for solving several theoretical and practical problems.

Wahed and Bishop presuppose $2n$ distinct eigenvalues, but in theory and practice one can find cases with multiple eigenvalues and an insufficient number of eigenvectors (the algebraic multiplicity of the eigenvalue is greater than its geometrical multiplicity; see Examples). Although this case is well known, e.g. [8, 9], it is desirable to extend Wahed's and Bishop's method to obtain the above mentioned advantages. The present paper evolves this extension. All essential features in the theory can be developed by considering the case of an eigenvalue with e.g. multiplicity three, having only one eigenvector, and this approach simplifies the demonstration of the method.

Finally it should be emphasised that Wahed's and Bishop's method is quite adequate for problems with an eigenvalue of multiplicity n having n independent eigenvectors.

2. Free Vibration

The equation of motion is in this case

$$A\ddot{q} + B\dot{q} + Cq = 0\,. \tag{2}$$

Seeking a solution of the form

$$q = \sigma\,\mathrm{e}^{\lambda t}\,, \tag{3}$$

we get the characteristic equation

$$|\lambda^2 A + \lambda B + C| = 0\,. \tag{4}$$

This is a polynomial of degree $2n$ in λ, provided that the mass matrix A is not singular, i.e. $|A| \neq 0$ and the inverse matrix A^{-1} exists. This assumption does not reduce generality very much, because according to the Lagrange formalism the quadratic expression in $\dot{q}$

$$T = \tfrac{1}{2}\,\dot{q}^T A \dot{q}$$

is the kinetic energy of the system. Therefore A is symmetric and positive definite. (Using the Newton-Euler formalism, one might get a nonsymmetric mass matrix; but after premultiplying the equation of motion with a suitable matrix, the new mass matrix is again symmetric. Furthermore in [10] a process of elimination is mentioned, which always leads to a symmetric mass matrix).

The $2n$ roots of (4) are the eigenvalues of the system and to each eigenvalue λ_r corresponds an eigenvector $\sigma^{(r)}$, satisfying

$$(\lambda_r^2 A + \lambda_r B + C)\,\sigma^{(r)} = 0 \tag{5}$$

and an eigenrow $\pi^{(r)}$, satisfying

$$\pi^{(r)}(\lambda_r^2 A + \lambda_r B + C) = 0\,. \tag{6}$$

Each $\sigma^{(r)}$ and each $\pi^{(r)}$ is determined except for an arbitrary multiplier. We now presume that besides the distinct eigenvalues λ_r ($r = 1, \ldots, 2n-3$) there is an eigenvalue λ with multiplicity three, to which there corresponds only one eigenvector σ, containing one arbitrary multiplier. Corresponding to λ, three independent solutions to (2) should be found. Formula (3) only gives one of them. In this case it is well known that further solutions can b) written in the form

$$q = \left(D + Et + F\frac{t^2}{2!}\right)e^{\lambda t}\,. \tag{7}$$

On substituting (7) we find that D, E an F have to fulfil

$$\begin{aligned} (\lambda^2 A + \lambda B + C)\,F &= 0\,,\\ (2\lambda A + B)\,F + (\lambda^2 A + \lambda B + C)\,E &= 0\,,\\ AF + (2\lambda A + B)\,E + (\lambda^2 A + \lambda B + C)\,D &= 0\,. \end{aligned} \tag{8}$$

Notice that F is the already mentioned eigenvector, containing one arbitrary constant f. E will contain one more arbitrary constant e besides, and D possesses three arbitraries f, e and d.

Now we can define the three *generalized eigenvectors* corresponding to the eigenvalue λ:

$$\sigma = D + Et + F\frac{t^2}{2!}, \qquad \dot\sigma = E + Ft\,, \qquad \ddot\sigma = F \tag{9}$$

and because of (8) these polynomials will satisfy the conditions:

$$\begin{aligned} (\lambda^2 A + \lambda B + C)\,\ddot\sigma &= 0\,,\\ (\lambda A + B)\,\ddot\sigma + (\lambda^2 A + \lambda B + C)\,\dot\sigma &= 0\,,\\ A\ddot\sigma + (\lambda A + B)\,\dot\sigma + (\lambda^2 A + \lambda B + C)\,\sigma &= 0\,. \end{aligned} \tag{10}$$

Introducing a modal matrix

$$\Sigma = [\sigma^{(1)} \ldots \sigma^{(2n-3)}\ \ddot\sigma\dot\sigma\sigma] \tag{11}$$

and a matrix

$$\Lambda = \operatorname{diag}\,[\lambda_1, \ldots \lambda_{2n-3}, \lambda, \lambda, \lambda, \lambda]\,, \tag{12}$$

(5) and (10) can be written in matrixform in the *eigenvector equation*:

$$A\ddot\Sigma + 2A\dot\Sigma\Lambda + B\dot\Sigma + A\Sigma\Lambda^2 + B\Sigma\Lambda + C\Sigma = 0\,. \tag{13}$$

The complete solution of equation (2) containing $2n$ arbitrary constants can be written as

$$q = \Sigma K\,, \tag{14}$$

where

$$K = \operatorname{diag}\,[e^{\lambda_1 t}, \ldots, e^{\lambda_{2n-3} t}, e^{\lambda t}, e^{\lambda t}, e^{\lambda t}]\,. \tag{15}$$

It is obvious how to define the generalized eigenrows corresponding to λ. Let F_1, E_1 and D_1 be rows of order $1 \times n$ satisfying

$$\begin{aligned} F_1(\lambda^2 A + \lambda B + C) &= 0\,,\\ F_1(2\lambda A + B) + E_1(\lambda^2 A + \lambda B + C) &= 0\,,\\ F_1 A + E_1(2\lambda A + B) + D_1(\lambda^2 A + \lambda B + C) &= 0\,. \end{aligned} \tag{16}$$

The eigenrow F_1 contains one arbitrary constant f_1, E_1 two arbitrary constants f_1 and e_1, and D_1 the three arbitraries f_1, e_1 and d_1.

Defining the *generalized eigenrows* as

$$\pi = D_1 - E_1 t + F_1\frac{t^2}{2!}, \qquad -\dot\pi = E_1 - F_1 t\,, \qquad \ddot\pi = F_1 \tag{17}$$

it is easy to show with help of (16) that these rows also satisfy

$$\begin{aligned} \ddot\pi(\lambda^2 A + \lambda B + C) &= 0\,,\\ \ddot\pi(2\lambda A + B) + (-\dot\pi)\,(\lambda^2 A + \lambda B + C) &= 0\,,\\ \ddot\pi A + (-\dot\pi)\,(2\lambda A + B) + \pi(\lambda^2 A + \lambda B + C) &= 0\,. \end{aligned} \tag{18}$$

The apparently unmotivated minus sign in (17) and (18) will be explained later.

With the modal matrix

$$\Pi = \begin{bmatrix} \pi^{(1)} \\ \vdots \\ \pi^{(2n-3)} \\ \pi \\ -\dot{\pi} \\ \ddot{\pi} \end{bmatrix} \tag{19}$$

(6) and (18) can be written in the *eigenrow equation*:

$$\ddot{\Pi}A - 2\Lambda\dot{\Pi}A - \dot{\Pi}B + \Lambda^2\Pi A + \Lambda\Pi B + \Pi C = 0 . \tag{20}$$

Finally we should briefly explain how to alter all definitions and equations in the case where the eigenvalue λ has a multiplicity $m = 1$, $m = 2$ or $m \geqq 4$.

For $m = 1$ is $\dot{\Sigma} = \ddot{\Sigma} = 0$, $\dot{\Pi} = \ddot{\Pi} = 0$ and the simplified forms of (13) and (20) can be found in [4].

For $m = 2$ all equations remain valid by assuming $\ddot{\sigma} = F = 0$ and $\ddot{\pi} = F_1 = 0$.

For $m = 4$ we have to replace (7) by

$$\sigma = D + Et + F\frac{t^2}{2!} + G\frac{t^3}{3!}$$

and the modal matrix Σ in (11) becomes

$$\Sigma = [\sigma^{(1)} \dots \sigma^{(2n-4)}\dddot{\sigma}\ddot{\sigma}\dot{\sigma}\sigma] .$$

The eigenvector equation (13) is valid as before. (17) must be replaced by

$$\pi = D_1 - E_1t + F_1\frac{t^2}{2!} - G_1\frac{t^3}{3!}, \qquad -\dot{\pi} = E_1 - F_1t + G_1\frac{t^2}{2!}, \qquad \ddot{\pi} = F_1 - G_1t , \qquad -\dddot{\pi} = G_1 \tag{21}$$

and these are the new elements in the modal matrix Π, while the eigenrow equation (20) is still valid. Note again the sign$(-1)^n$ before the term t^n when defining the generalized eigenrows by (21).

3. Forced Vibration

As with conventional modal analysis we seek to write the solution of

$$A\ddot{q} + B\dot{q} + Cq = Q(t) \tag{22}$$

in the form

$$q = \chi_1(t)\,\sigma^{(1)} + \dots + \chi_{2n-3}(t)\,\sigma^{(2n-3)} + \chi_{2n-2}(t)\,\ddot{\sigma} + \chi_{2n-1}(t)\,\dot{\sigma} + \chi_{2n}(t)\,\sigma , \tag{23}$$

where $\chi_i(t)$ are functions of time.

For reasons of matrix formulation we introduce

$$\chi = \begin{bmatrix} \chi_1(t) \\ \vdots \\ \chi_{2n}(t) \end{bmatrix} \tag{24}$$

and using the definitions (11), (12), (19) and (24) we state the

Theorem: *Equation* (22) *has a solution*

$$q = \Sigma\chi \tag{25}$$

with the restriction

$$\Sigma\dot{\chi} = \Sigma\Lambda\chi . \tag{26}$$

The matrix χ can be found by solving

$$\dot{\chi} - \Lambda\chi = \Pi Q(t) , \tag{27}$$

which represents 2n uncoupled differential equations of first order for the functions of time $\chi_i(t)$.

The modal matrices Σ and Π must be normalized by

$$\Sigma\Pi = 0 \tag{28}$$

and

$$\Sigma\Lambda\Pi + \dot{\Sigma}\Pi = A^{-1} . \tag{29}$$

It is worth noting that expression (25) represents the complete solution to equation (22) in the case, we choose to solve (27) with $2n$ constants of integration.

Before proving the theorem, first a couple of remarks concerning the normalization conditions (28) and (29).

They were proved in the case of single eigenvalues ($\dot{\Sigma} = 0$) in [4] in a complicated way as a property of the matrices Σ and Π. Here we want to understand them as conditions for Σ and Π and will only show how natural it is to assert them. With help of the unit matrix I, (26) can be written as

$$\Sigma(I\dot{\chi} - \Lambda\chi) = 0 . \tag{30}$$

On substituting (27) into (30) we get

$$\Sigma\Pi Q(t) = 0 . \tag{31}$$

Because this must be valid for all matrices $Q(t)$, the necessity of normalization equation (28) is quite evident. In addition equation (28) implies

$$\dot{\Sigma}\Pi + \Sigma\dot{\Pi} = 0 , \tag{32}$$

which with (11) and (19) means that

$$\ddot{\sigma}\dot{\pi} + \ddot{\sigma}(-\dot{\pi}) + \dot{\sigma}(-\ddot{\pi}) + \dot{\sigma}\ddot{\pi} = 0 , \tag{33}$$

which is obviously right.

Here we have the reason for choosing the sign $(-1)^n$ before the term t^n when defining the generalized eigenrows in (16), (17) and (21).

To give the reasons for normalization equation (29) is a bit more complicated. We postmultiply the eigenvector equation (13) by ΠA and premultiply the eigenrow equation (20) by $A\Sigma$. Subtraction of these two new equations together with equation (28), gives

$$2A\dot{P}B + BPA - APB = 0 . \tag{34}$$

Here we have used the simplification

$$P = \Sigma\Lambda\Pi + \dot{\Sigma}\Pi . \tag{35}$$

With help of (11), (12) and (19) we see that

$$\dot{P} = 0 . \tag{36}$$

Substituting (36) in (34), we get a condition for the matrix P:

$$BPA = APB . \tag{37}$$

This equation can of course have many solutions P. We can mention for example A^{-1} or B^{-1}. But the matrices A and B were assumed to be totally arbitrary except for the existence of A^{-1}. So it is natural to claim $P = A^{-1}$ which gives equation (29).

Finally the normalization equations (28) and (29) imply for distinct eigenvalues λ_r the normalization [1, 4, 5]

$$\pi^{(r)}(2\lambda_r A + B)\,\sigma^{(r)} = 1 \tag{39}$$

and for the eigenvalue λ with multiplicity three a new normalization

$$\ddot{\pi}(2\lambda A + B)\,\sigma + \ddot{\pi}A\dot{\sigma} = 1 . \tag{40}$$

This can be seen with help of equation

$$H\Pi = (\Lambda\Pi A\Sigma + \Pi A\Sigma\Lambda + \Pi A\dot{\Sigma} + \Pi B\Sigma)\,\Pi = \Pi . \tag{41}$$

This equation is evident assuming (28) and (29).

In the case where all eigenvalues are distinct, one can claim as normalization the matrix H to be the unit matrix I and this leads to equation (39). In our case this is no longer possible. But analysing the matrix H in detail, the requirement (39) for distinct eigenvalues can be maintained and requirement (40) for multiple eigenvalues becomes suitable.

Now we have to prove the theorem.

Assuming

$$q = \Sigma\chi \tag{25}$$

we have

$$\dot{q} = \Sigma\dot{\chi} + \dot{\Sigma}\chi \tag{42}$$

and

$$\ddot{q} = \Sigma\ddot{\chi} + 2\dot{\Sigma}\dot{\chi} + \ddot{\Sigma}\chi . \tag{43}$$

From equation (27) we get

$$\dot{\chi} = \Pi Q + \Lambda\chi \tag{44}$$

and

$$\ddot{\chi} = \dot{\Pi}Q + \Pi\dot{Q} + \Lambda(\Pi Q + \Lambda\chi) . \tag{45}$$

On substituting (44) and (45) into (42) and (43) we get

$$\dot{q} = \Sigma\Pi Q + \Sigma\Lambda\chi + \dot{\Sigma}\chi \tag{46}$$

and

$$\ddot{q} = \Sigma\dot{\Pi}Q + \Sigma\Pi\dot{Q} + \Sigma\Lambda\Pi Q + \Sigma\Lambda^2\chi + 2\dot{\Sigma}\Pi Q + 2\dot{\Sigma}\Lambda\chi + \ddot{\Sigma}\Pi . \tag{47}$$

Now we write the left side of the differential equation (22) with help of (46) and (47):

$$A\ddot{q} + B\dot{q} + Cq = A(\Sigma\dot{\Pi}Q + \Sigma\Pi\dot{Q} + \Sigma\Lambda\Pi Q + \Sigma\Lambda^2\chi + 2\dot{\Sigma}\Pi Q + 2\dot{\Sigma}\Lambda\chi + \ddot{\Sigma}\Pi) + \\ + B(\Sigma\Pi Q + \Sigma\Lambda\chi + \dot{\Sigma}\chi) + C\Sigma\chi . \tag{48}$$

Several terms on the right side of equation (48) vanish because of the eigenvector equation (13) and the normalization equation (28). So equation (48) simplifies to

$$A\ddot{q} + B\dot{q} + Cq = A(\Sigma\dot{\Pi} + \Sigma\Lambda\Pi + 2\dot{\Sigma}\Pi)\,Q\,. \tag{49}$$

Using (32) we can also write

$$A\ddot{q} + B\dot{q} + Cq = A(\Sigma\Lambda\Pi + \Sigma\dot{\Pi})\,Q\,. \tag{50}$$

Now the second normalization equation (29) can be applied and (50) becomes

$$A\ddot{q} + B\dot{q} + Cq = Q\,,$$

which shows the correctness of the theorem.

4. Harmonic Excitation

When the system is subjected to harmonic excitation, the equation of motion is

$$A\ddot{q} + B\dot{q} + Cq = \Phi\,\mathrm{e}^{i\omega t}\,, \tag{51}$$

in which Φ is a constant column.

Applying the theorem from before, equation (27) becomes

$$\dot{\chi} - \Lambda\chi = \Pi\Phi\,\mathrm{e}^{i\omega t}\,. \tag{52}$$

Remember that the modal matrix Π in our case of multiple eigenvalues contains polynomials in time.

We try to find a solution to (52) in the form

$$\chi = (i\omega I - \Lambda)^{-1}\,\Pi\Phi\,\mathrm{e}^{i\omega t} + R(t)\,, \tag{53}$$

where $R(t)$ contains n time functions ordered in a column. Substituting (53) into (52) we see that $R(t)$ has to satisfy the equation

$$\dot{R} - \Lambda R = (i\omega I - \Lambda)^{-1}\,(-\dot{\Pi})\,\Phi\,\mathrm{e}^{i\omega t}\,. \tag{54}$$

As before we seek a solution R of equation (54) in the form

$$R = (i\omega I - \Lambda)^{-2}\,(-\dot{\Pi})\,\Phi\,\mathrm{e}^{i\omega t} + S(t) \tag{55}$$

and it is easily shown that S has to satisfy the equation

$$\dot{S} - \Lambda S = (i\omega I - \Lambda)^{-2}\,\ddot{\Pi}\Phi\,\mathrm{e}^{i\omega t}\,. \tag{56}$$

The matrix $\ddot{\Pi}$ is constant, because the highest multiplicity of an eigenvalue is assumed to be three, and therefore a solution to (56) can be written immediately:

$$S = (i\omega I - \Lambda)^{-3}\,\ddot{\Pi}\Phi\,\mathrm{e}^{i\omega t}\,. \tag{57}$$

Substituting (57) into (55) and (55) into (53) the solution of (52) becomes

$$\chi = \{(i\omega I - \Lambda)^{-1}\,\Pi + (i\omega I - \Lambda)^{-2}\,(-\dot{\Pi}) + (i\omega I - \Lambda)^{-3}\,\ddot{\Pi}\}\,\Phi\,\mathrm{e}^{i\omega t}\,. \tag{58}$$

Using the normalization equations (28) and (29) a solution to (51) becomes

$$q = \Sigma\chi = \Sigma[(i\omega I - \Lambda)^{-1}\,\Pi + (i\omega I - \Lambda)^{-2}\,(-\dot{\Pi}) + (i\omega I - \Lambda)^{-3}\,\ddot{\Pi}]\,\Phi\,\mathrm{e}^{i\omega t}\,. \tag{59}$$

If all eigenvalues are distinct, we have $-\dot{\Pi} = \ddot{\Pi} = 0$, and result (59) is then equivalent to that in [1.4].

Formula (59) can with (11) and (19) also be written as

$$q = \left[\sum_{r=1}^{2n-3} \frac{\sigma^{(r)}\pi^{(r)}}{i\omega - \lambda_r} + \frac{\ddot{\sigma}\pi + \dot{\sigma}(-\dot{\pi}) + \sigma\ddot{\pi}}{i\omega - \lambda} + \frac{\ddot{\sigma}(-\dot{\pi}) + \dot{\sigma}\ddot{\pi}}{(i\omega - \lambda)^2} + \frac{\ddot{\sigma}\ddot{\pi}}{(i\omega - \lambda)^3}\right]\Phi\,\mathrm{e}^{i\omega t}. \tag{60}$$

This expression is the steady-state solution of (51), because the differentiation of the bracket in (60) results in zero.

5. Examples

First we consider the system

$$\begin{bmatrix}1 & 0\\ 0 & 1\end{bmatrix}\ddot{q} + \begin{bmatrix}-1 & 0\\ 1 & 3\end{bmatrix}\dot{q} + \begin{bmatrix}1 & 3\\ 1 & 2\end{bmatrix}q = Q(t)\,. \tag{61}$$

The characteristic equation (4) gives the eigenvalue $\lambda_1 = 1$ and the eigenvalue $\lambda = -1$ with multiplicity three. According to (5) and (6), corresponding to λ_1 we have the eigenvector $\sigma^{(1)} = \begin{bmatrix}-3\alpha\\ \alpha\end{bmatrix}$ and the eigenrow $\pi^{(1)} = [-2\beta\ \beta]$.

Corresponding to λ we can only find one eigenvector $\begin{bmatrix}-f\\ f\end{bmatrix}$ and one eigenrow $[0\ f_1]$.

The generalized eigenvectors (9) can now be determined by (8) and the generalized eigenrows (17) by (16). The modal matrices (11) and (19) then become

$$\Sigma = [\sigma^{(1)}\ddot{\sigma}\dot{\sigma}\sigma] = \begin{bmatrix} -3\alpha & -f & -f - e - ft & -\frac{2}{3}f - e - d - ft - ef - t\frac{t^2}{2} \\ \alpha & f & e + ft & d + et + f\frac{t^2}{2} \end{bmatrix}, \tag{62}$$

$$\Pi = \begin{bmatrix} \pi^{(1)} \\ \pi \\ -\dot{\pi} \\ \ddot{\pi} \end{bmatrix} = \begin{bmatrix} -2\beta & \beta \\ -\frac{f_1}{3} - \frac{e_1}{3} + \frac{f_1}{3}t & d_1 - e_1 t + f_1 \frac{t^2}{2} \\ -\frac{f_1}{3} & e_1 - f_1 t \\ 0 & f_1 \end{bmatrix}. \tag{63}$$

We can now with advantage use the normalization equations (39) and (40) and they give the relations

$$\alpha\beta = \tfrac{1}{8}, \qquad f f_1 = -\tfrac{3}{2}. \tag{64}$$

The two missing normalization relations are obtained from (28) and (29) (which of course also give the results in (64); this shows that the use of (39) and (40) is not necessary, merely convenient).

In our case (28) and (29) take the following form:

$$\sigma^{(1)}\pi^{(1)} + \ddot{\sigma}\pi + \dot{\sigma}(-\dot{\pi}) + \sigma\ddot{\pi} = 0, \tag{28}$$

$$\lambda_1\sigma^{(1)}\pi^{(1)} + \lambda\ddot{\sigma}\pi + \lambda\dot{\sigma}(-\dot{\pi}) + \lambda\sigma\ddot{\pi} + \ddot{\sigma}(-\dot{\pi}) + \dot{\sigma}\ddot{\pi} = A^{-1}. \tag{29}$$

Therefrom we find

$$f e_1 + e f_1 = \tfrac{3}{4} \quad \text{and} \quad f d_1 + e e_1 + d f_1 = -\tfrac{1}{8}. \tag{65}$$

According to (64) and (65) we choose for example

$$\alpha = 1, \quad \beta = \tfrac{1}{8}, \quad f = 1, \quad f_1 = -\tfrac{3}{2}, \quad e = 0, \quad e_1 = \tfrac{3}{4}, \quad d = 0, \quad d_1 = -\tfrac{1}{8}. \tag{66}$$

The values for α, β, f, e, d, f_1, e_1 and d_1 should be substituted into (62) and (63).

Writing $Q(t) = \begin{bmatrix} Q_1(t) \\ Q_2(t) \end{bmatrix}$, the differential equations (27) now become

$$\begin{aligned} \dot{\chi}_1 - \chi_1 &= -\tfrac{1}{4} Q_1(t) + \tfrac{1}{8} Q_2(t), \\ \dot{\chi}_2 + \chi_2 &= (\tfrac{1}{4} - \tfrac{1}{2}t)\, Q_1(t) - (\tfrac{1}{8} + \tfrac{3}{4}t + \tfrac{3}{4}t^2)\, Q_2(t). \\ \dot{\chi}_3 + \chi_3 &= \tfrac{1}{2} Q_1(t) + (\tfrac{3}{4} + \tfrac{3}{2}t)\, Q_2(t), \\ \dot{\chi}_4 + \chi_4 &= -\tfrac{3}{2} Q_2(t). \end{aligned} \tag{67}$$

Choosing concrete functions $Q_1(t)$ and $Q_2(t)$, equations (67) can be solved and the results $\chi_1(t), \dots, \chi_4(t)$ together with (62) can be used to write the solution to (61) in the form (25):

$$q = \Sigma\chi = \chi_1(t)\,\sigma^{(1)} + \chi_2(t)\,\ddot{\sigma} + \chi_3(t)\,\dot{\sigma} + \chi_4(t)\,\sigma. \tag{68}$$

As another simple example we can mention

$$\begin{bmatrix} 5 & 2 \\ 2 & 1 \end{bmatrix} \begin{bmatrix} \ddot{\theta}_1 \\ \ddot{\theta}_2 \end{bmatrix} + \begin{bmatrix} -2 & 4 \\ -2 & 2 \end{bmatrix} \begin{bmatrix} \theta_1 \\ \theta_2 \end{bmatrix} = \begin{bmatrix} e^{2it} \\ 0 \end{bmatrix} \tag{69}$$

which represents the equations of motion, [11], for the system shown in figure 1, assuming additionally an external torque $M = e^{2it}$ on the bar AB. (69) has two eigenvalues, both with multiplicity two and both possessing only one eigenvector. The steady-state solution can be found by section 4 to be

$$\begin{bmatrix} \theta_1 \\ \theta_2 \end{bmatrix} = \begin{bmatrix} -\tfrac{1}{2}\, e^{i2t} \\ \tfrac{5}{2}\, e^{i2t} \end{bmatrix}, \tag{70}$$

a result which can easily be seen to be correct by substituting it into the equation of motion (69).

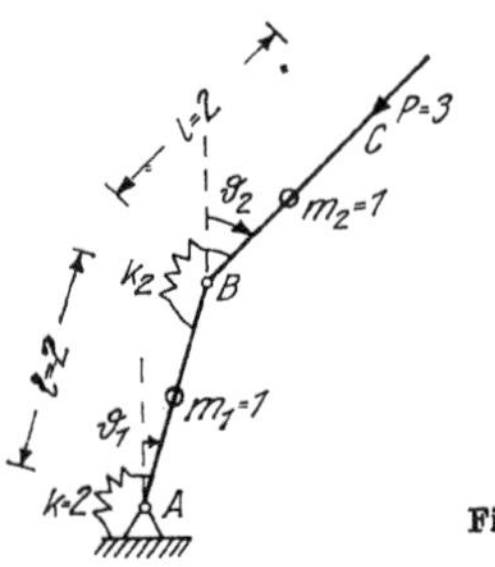

Fig. 1

6. Conclusions

A linear non-conservative system can possess multiple eigenvalues with an insufficient number of eigenvectors. In this paper a theory for such systems is developed, which permits calculation of the system's response to harmonic and to general excitation. In all essential the theory is an extension of a work by WAHED and BISHOP (1), which is limited to distinct eigenvalues. In their case, eigenvectors and eigenrows contain constants; for multiple eigenvalues an obvious idea is to use polynomials in time. A little difficulty then arises as to the sign in the eigenrows:

$(-1)^n$ must be placed before all terms t^n. The modal matrices Σ and Π, consisting of the eigenvectors and eigenrows, should be normalized by the equations $\Sigma\Pi = 0$ and $\Sigma\Lambda\Pi + \dot{\Sigma}\Pi = A^{-1}$. Now the systems response to an excitation $Q(t)$ can be written as $q = \Sigma\chi$, where χ is the solution to $\dot{\chi} - \Lambda\chi = \Pi Q(t)$. If $Q(t)$ is chosen as a harmonic excitation $\Phi\, e^{i\omega t}$, the results in the theory resembles the well known expressions from modal analysis for conservative systems.

7. Appendix

It should be explained how the mass matrix A, the damping matrix B and the stiffness matrix C can be determined from the matrix Λ, containing all eigenvalues and the normalized modal matrices Σ and Π. WOODCOCK [2] and FAWZY/BISHOP [4] has already answered a similar question concerning distinct eigenvalues. First of all, the equation of normalization (29) gives the result

$$A = (\Sigma\Lambda\Pi + \dot{\Sigma}\Pi)^{-1} . \tag{71}$$

In addition, postmultiplying the eigenvector equation (13) by Π, we get:

$$A\Sigma\Lambda^2\Pi + B\Sigma\Lambda\Pi + C\Sigma\Pi + 2A\dot{\Sigma}\Lambda\Pi + B\dot{\Sigma}\Pi + A\ddot{\Sigma}\Pi = 0 . \tag{72}$$

Using both normalization equations (28) and (29), equation (72) becomes

$$A\Sigma\Lambda^2\Pi + BA^{-1} + 2A\dot{\Sigma}\Lambda\Pi + A\ddot{\Sigma}\Pi = 0 . \tag{73}$$

Postmultiplying by A, solving with respect to B and using (71) gives

$$B = -(\Sigma\Lambda\Pi + \dot{\Sigma}\Pi)^{-1} (\Sigma\Lambda^2 + 2\dot{\Sigma}\Lambda + \ddot{\Sigma})\, \Pi(\Sigma\Lambda\Pi + \dot{\Sigma}\Pi)^{-1} . \tag{74}$$

Concerning the determination of C, which leads to a rather long formula, we will merely mention that the eigenrow equation (20) may be premultiplied by $A\Sigma\Lambda + A\dot{\Sigma}$. With help of the normalization equation (29) and the results for A and B in (71) and (74), an expression for C can be found.

References

1 WAHED, I. F. A., BISHOP, R. E. D., On the equations governing the free and forced vibrations of a general non-conservative system, J. Mech. E. **18** (1976), 6–10.
2 WOODCOCK, D. L., On the interpretation of the vector plots of forced vibrations of a linear system with viscous damping, Aeronaut. Q. **14** (1963), 45–62.
3 SCHMIDTZ, P. D., Normal mode solution to the equations of motion of a flexible airplane, J. Aircr. **10** (1973), 318–320.
4 FAWZY, I., BISHOP, R. E. D., On the dynamics of linear non-conservative systems, Proc. R. Soc. Lond. A. **352** (1976), 25–40.
5 FAWZY, I., BISHOP, R. E. D., On the nature of resonance in non-conservative systems, Journal of Sound and Vibrations **55** (4) (1977), 475–485.
6 BISHOP, R. E. D., PRICE, W. G., An investigation into the linear theory of ship response to waves, Journal of Sound and Vibrations **62** (3) (1979), 353–363.
7 LUND, J. W., Response by normal coodinates of a structure interacting with a fluid, K. MAGNUS (Ed.), Dynamic of Multibody Systems, Springer, Berlin/Heidelberg/New York, 1978, 196–207.
8 BOYCE, W. E., DI PRIMA, R. C., Elementary differential equations, Wiley, 1969.
9 PIPES, L. A., Matrix methods for engineering, Prentice-Hall, Inc., 1963.
10 MÜLLER, P. C., SCHIEHLEN, W. O., Forced Linear Vibrations, Courses and Lectures, CISM, Udine, 16.
11 MÜLLER, P. C., Stabilität und Matrizen, Springer, Berlin/Heidelberg/New York, 1977, 49.

Eingereicht am 20. 4. 1980

Anschrift: Lektor WOLFHARD KLIEM, Danmarks Ingeniörakademi, Maskinafdelingen, Akademivej, bygning 358, 2800 Lyngby, Danmark.

ZAMM 60, 577 – 585 (1980)

Cz. M. Rodkiewicz / R. N. Gupta

The Weak Interaction Induced Pressure due to the Stepwise-Deceleration of a Flat Plate in Hypersonic Flow

Ein Plattenstreifen wird einer stufenförmig gebremsten Überschallströmung bei einem Angriffswinkel von 0° ausgesetzt. Die Grenzschichtkonfiguration führt zu einem induzierten Druck. Um diese zeitliche Druckverteilung zu erhalten, wurden entsprechend dem Vorgehen in (1) *die transformierten Impuls- und Energiegleichungen gelöst. Diese Lösungen werden zur Berechnung der Verschiebungsstärke der unstetigen Grenzschicht herangezogen. Die differenzierte Verschiebungsstärke wird schließlich für den Ausdruck des induzierten Druckes genutzt, wobei die akustische Näherung enthalten ist.*

A semi-infinite plate is exposed to the stepwise-decelerated hypersonic flow at zero angle of attack. The boundary layer reconfiguration gives rise to an induced pressure. To obtain this temporal pressure distribution, the transformed momentum and energy equations have been solved using the approach of Ref. [1]. *These solutions are used to evaluate the displacement thickness for the unsteady boundary layer. The differentiated displacement thickness is finally used in the induced pressure expression involving the acoustic approximation. The shearing stress and heat transfer at the wall are obtained.*

Полоса пластины подвергается действию замедленного ступенчатым образом сверхзвукового потока с нулевым углом атаки. Конфигурация пограничного слоя приводит к индуцированному давлению. Для получения распределения давления во времени решаются трансформированные уравнения моментов и энергии соответственно работе [1]. Эти решения используются для расчёта толщины перемещения прерывного пограничного слоя. Дифференцированная толщина перемещения используется в конце для выражения индуцированного давления содержащего звуковое приближение.

List of Symbols

a	velocity of sound
$a_{m,n}$	coefficient given by equation (34)
a	a function of g_w in equation (55)
$b_{m,n}$	coefficient given by equation (35)
C	a function of η and τ
C	$= C(X)$ or a constant
$c_{m,n}$	coefficient given by equation (36)
Cp	specific heat at constant pressure
$d_{m,n}$	coefficient given by equation (37)
$e_{m,n}$	coefficient given by equation (38)
f	function related to the stream function by $f(\eta, \tau) = \psi \sqrt{\left[\frac{1}{2\nu_e C x U_e}\right]}$
g	function related to enthalpy by $g = \frac{h}{h_e}$
h	enthalpy
H	stepsize in the η-direction
I	integral appearing in the displacement thickness
J	integral appearing in the displacement thickness
k	thermal conductivity
m	an integer in numerical method
M	Mach Number
n	an integer in numerical method
N	value of n for which "e" condition prevails
p	thermodynamic pressure
Pr	Prandtl Number
q	heat transfer per unit time per unit area
R	a function of η and τ
Re	Reynolds number
S	$\equiv \Delta f$ (stream function increment) used in the computer programming
t	time
T	absolute temperature
T	stepsize in the τ-direction
U	stream velocity outside the boundary layer in X-direction
u	velocity component parallel to wall
v	velocity component normal to wall
x	Cartesian co-ordinate parallel to wall
y	Cartesian co-ordinate normal to wall
Y	modified co-ordinate in y-direction
α_1	as defined by equation (60)
α_2	as defined by equation (62)
γ	ratio of specific heats
δ^*	displacement thickness as defined by equation (45)
δ_ϱ	as defined by equation (46)
Δ	displacement height (unsteady)
η	similarity independent variable associated with space
λ	as defined by equation (55)
μ	dynamic viscosity coefficient
ν	kinematic viscosity coefficient
ϱ	density
τ	shear stress

τ similarity independent variable associated with space and time
$\bar{\chi}$ $\equiv \chi$ hypersonic similarity parameter defined by expression (53)
ψ stream function

Subscripts

w evaluation at wall ($y = 0$)
e evaluation in free stream ($y = y_e$)
1 initial steady state
2 final steady state

I. Introduction

The time-dependent viscous interaction plays an important role in the evaluation of the forces acting on the components of the aerodynamic vehicles maneuvering at Hypersonic speeds. The unsteady pressure problem for the case of a stepwise-accelerated flat plate has been analysed in reference [1]. The magnitude of the transient contribution to the induced pressure is obtained therein by an investigation of the time-dependent boundary layer flow. It has been suggested that

$$\eta = \left[\frac{U_e}{2\nu_e C x}\right] Y\,, \quad \text{and} \quad \tau = \frac{U_e t}{x}$$

are the proper independent variables to transform the non-steady boundary layer equations to similarity forms for the case of zero pressure gradient.

In the present analysis consideration is given to the stepwise-decelerated flat plate in Hypersonic Flow. The solutions incorporate a wide range of Mach Number and wall heat transfer (including the adiabatic wall) for Prandtl Number Unity.

II. Time-dependent Compressible Boundary Layer Equations in Two-Dimensions

Prandtl's Boundary Layer Equations in the unsteady form are taken as the equations governing the flow in a time-dependent compressible, laminar, two-dimensional boundary layer. They are [2], for the case of zero pressure gradient,

$$\varrho\left(\frac{\partial u}{\partial t} + u\frac{\partial u}{\partial x} + v\frac{\partial u}{\partial y}\right) = \frac{\partial}{\partial y}\left(\mu\frac{\partial u}{\partial y}\right) \tag{1}$$

$$0 = \frac{\partial p}{\partial y}, \tag{2}$$

$$\varrho\left(\frac{\partial h}{\partial t} + u\frac{\partial h}{\partial x} + v\frac{\partial h}{\partial y}\right) = \frac{\partial p}{\partial t} + \frac{\partial}{\partial y}\left(\frac{\mu}{Pr}\frac{\partial h}{\partial y}\right) + \mu\left(\frac{\partial u}{\partial y}\right)^2 \tag{3}$$

$$\frac{\partial \varrho}{\partial t} + \frac{\partial}{\partial x}(\varrho u) + \frac{\partial}{\partial y}(\varrho v) = 0 \tag{4}$$

with the boundary conditions:

$$\begin{aligned} &\text{when } y = 0 && u = v = 0\,, \\ &&& h = h_w = \text{const.} \\ &\text{when } y = y_e && u = U_e = \text{const.} \\ &&& h = h_e = \text{const.} \end{aligned} \tag{5}$$

The pressure in the boundary layer being independent of y, may be given by its value at $y = y_e$ according to the inviscid solution

$$\varrho_e\left(\frac{\partial U_e}{\partial t} + U_e\frac{\partial U_e}{\partial x}\right) = \frac{\partial p}{\partial x} = 0 \tag{6}$$

where U_e denotes the prescribed potential motion.

III. Reduction to Similarity Forms

Using the following forms of the similarity independent variables,

$$\eta = \sqrt{\frac{U_e}{2\nu_e C x}}\, Y\,, \tag{7}$$

$$\tau = \frac{U_e t}{x}, \tag{8}$$

equations (1) and (3) reduce to, respectively, for the case of weak interactions (initial zero pressure gradient):

$$2\dot{f}' + 2\tau(\dot{f}f'' - \dot{f}'f') - f''' - ff'' = 0\,; \tag{9}$$

$$2Pr(1 - \tau f')\,\dot{g} - g'' - Pr(f - 2\tau\dot{f})\,g' - Pr(\gamma - 1)\,M_e^2 f''^2 = 0 \tag{10}$$

where the following relations have been used

$$Y = \int_0^y \frac{\varrho}{\varrho_e}\,dy\,, \qquad \frac{\partial\psi}{\partial y} = \frac{\varrho}{\varrho_e}u; \quad u = \frac{\partial\psi}{\partial Y}, \qquad v = -\frac{\varrho_e}{\varrho}\left(\frac{\partial\psi}{\partial x} + \frac{\partial Y}{\partial x}\frac{\partial\psi}{\partial Y} + \frac{\partial Y}{\partial t}\right) \tag{11), (12), (13}$$

$$C \equiv C(x) = \frac{\mu\varrho}{\mu_e\varrho_e}, \qquad U_e = \text{const.}\,, \qquad \psi = \sqrt{2\nu_e C x U_e}\,f(\eta,\tau)\,, \qquad h = g(\eta,\tau)\,h_e\,. \tag{14), (15), (16), (17}$$

Assuming that the free stream adjusts itself instantaneously to the new conditions, the following relations have also been used:

$$\frac{\partial U_e}{\partial t} = \frac{\partial h_e}{\partial t} = 0\,. \tag{18}$$

IV. Linearized Form of the Problem

Equation (9) may be linearized in a way similar to that of reference [1].

Let us assume that

$$f(\eta,\tau) = f_2(\eta) - \Delta f(\eta,\tau) \tag{19}$$

where $f_2(\eta)$ satisfies the steady state equation, namely,

$$f_2''' + f_2 f_2'' = 0\,. \tag{20}$$

Also, the steady state energy equation is

$$g_2'' + Pr f_2 g_2' + Pr(\gamma - 1)\,M_{e2}^2 f_2''^2 = 0\,. \tag{21}$$

Substitution of expression (19) into equation (9) results in

$$2(\dot{\Delta f})' + 2\tau[f_2''(\dot{\Delta f}) - f_2'(\dot{\Delta f})'] - (\Delta f)''' - f_2(\Delta f)'' - f_2''(\Delta f) = 0 \tag{22}$$

where the second order terms in Δf have been dropped.

V. Associated Boundary and Time Conditions

$$y = 0\,, \quad Y = 0\,, \quad \eta = 0\,, \quad u = 0\,, \quad f'(0,\tau) = 0\,, \quad f_2'(0) = 0\,, \quad (\Delta f)'(0,\tau) = 0; \tag{23}$$

$$y = 0\,, \quad Y = 0\,, \quad \eta = 0\,, \quad \psi = 0\,, \quad f(0,\tau) = 0\,, \quad f_2(0) = 0\,, \quad \Delta f(0,\tau) = 0; \tag{24}$$

$$y = y_e\,, \quad Y = Y_e\,, \quad \eta = \eta_e\,, \quad u = U_{e2}\,, \quad f'(\eta_e,\tau) = 1\,, \quad f_2'(\eta_e) = 1\,, \quad (\Delta f)'(\eta_e,\tau) = 0; \tag{25}$$

$$t = 0\,, \quad \tau = 0\,, \quad u = u_1 - (U_{e1} - U_{e2}), \quad f'(\eta > 0, 0) = f_1'\frac{U_{e1}}{U_{e2}} - \left(\frac{U_{e1}}{U_{e2}} - 1\right),$$

and

$$(\Delta f)'(\eta > 0, 0) = -f_1'\frac{U_{e1}}{U_{e2}} + \left(\frac{U_{e1}}{U_{e2}} - 1\right) + f_2'(\eta)\,. \tag{26}$$

The difference between the present problem and that of reference [1] arises essentially through this initial condition:

$$y = 0\,, \quad Y = 0\,, \quad \eta = 0\,, \quad h = h_w\,, \quad g(0,\tau) = \frac{h_w}{h_{e2}}, \tag{27}$$

and for the case of an insulated plate

$$g'(0,\tau) = 0; \tag{28}$$

$$y = y_e\,, \quad Y = Y_e\,, \quad \eta = \eta_e\,, \quad h = h_{e2}\,, \quad g(\eta_e,\tau) = 1; \tag{29}$$

$$t = 0\,, \quad \tau = 0\,, \quad h = h_1\,, \quad g(\eta, 0) = \frac{h_1}{h_{e2}}\,. \tag{30}$$

VI. Numerical Solution to the Momentum Equation

The solution to equation (22) with the time-dependent boundary conditions (23) through (26) has been obtained. The following recurrence relation of reference [1] has been used.

$$S_{m,n} = R_{m,n}S_{m,n+1} + C_{m,n} \tag{31}$$

where

$$R_{m,n} = \frac{-a_{m,n}}{b_{m,n} + C_{m,n}R_{m,n-1} + d_{m,n}R_{m,n-1}R_{m,n-2}} \tag{32}$$

and

$$C_{m,n} = \frac{e_{m,n} - (c_{m,n}C_{m,n-1} + d_{m,n}R_{m,n-2}C_{m,n-1} + d_{m,n}C_{m,n-2})}{b_{m,n} + c_{m,n}R_{m,n-1} + d_{m,n}R_{m,n-1}R_{m,n-2}} \tag{33}$$

and where $R_{m,1} = C_{m,1} = C_{m,2} = 0$, $R_{m,2} = 0.25$ are obtained from the boundary conditions (23) and (24).

Also, $a_{m,n}$, $b_{m,n}$, $c_{m,n}$, $d_{m,n}$ and $e_{m,n}$ are defined by the following equations

$$a_{m,n} = T + HTf_2(m, n) - H^2[1 - \tau_m f_2'(m, n)], \tag{34}$$

$$b_{m,n} = -3T - 2HTf_2(m, n) + H^3 T f_2''(m, n) - 2H^3 \tau_m f_2''(m, n), \tag{35}$$

$$c_{m,n} = 3T + HTf_2(m, n) + H^2[1 - \tau_m f_2'(m, n)], \tag{36}$$

$$d_{m,n} = -T, \tag{37}$$

$$e_{m,n} = H^2[1 - \tau_m f_2'(m, n)] [S_{m-1,n-1} - S_{m-1,n+1}] - 2H^3 \tau_m f_2''(m, n) S_{m-1,n}. \tag{38}$$

For $n = N$, where the 'e' condition prevails, the following relation may be used in view of the boundary condition (25):

$$S_{m,N} = \frac{C_{m,N-1} - \dfrac{C_{m,N-2}}{4 - R_{m,N-2}}}{\dfrac{3}{4 - R_{m,N-2}} - R_{m,N-1}}. \tag{39}$$

In reference [1], the Δf-distribution was obtained up to $\tau = 1$. In the present work, the Δf-distribution has been obtained up to $\tau = 1.3$. This is achieved by the use of double precisión in the computational technique.

Some efforts were also made to use the 11-point computational molecule in place of the 7-point computational molecule. The 11-point molecule is the result of the commonly used expressions for central derivatives. It was found that the numerical methods incorporating the 11-point molecule are not stable for the problem under consideration.

VII. Solution to the Energy Equation

The following solutions, for PRANDTL Number unity may be used [1]:

(a) For the case of no wall heat transfer

$$g = 1 + \frac{\gamma - 1}{2} M_e^2(1 - f'^2). \tag{40}$$

(b) For the case of non-zero wall heat transfer

$$g = 1 + \frac{\gamma - 1}{2} M_e^2(f' - f'^2) + (g_w - 1)(1 - f') \tag{41}$$

provided that the wall temperature is kept constant at all the times.

VIII. Weak Viscous Interactions

In the acoustic approximation, the pressure rise at the plate surface and the outward velocity of the "displacement plane" may be written [3] as, respectively,

$$\Delta p_v = \varrho_e a_e v_e \tag{42}$$

and

$$v_e = \frac{D\Delta}{Dt} = \frac{\partial \Delta}{\partial t} + U_e \frac{\partial \Delta}{\partial x}. \tag{43}$$

The displacement thickness for the time-dependent boundary layer, with constant ϱ_e and constant U_e, may be obtained from reference [4] as

$$\frac{\partial}{\partial t}(\Delta - \delta_\varrho) + U_e \frac{\partial}{\partial x}(\Delta - \delta^*) = 0 \tag{44}$$

where

$$\delta^* = \sqrt{\frac{2\nu_e C x}{U_e}} \int_0^e \left[g_w + \frac{(\gamma - 1)}{2} M_e^2 f'\right](1 - f')\, d\eta \tag{45}$$

and

$$\delta_\varrho = \sqrt{\frac{2\nu_e C x}{U_e}} \int_0^e \left\{\left[g_w + \frac{(\gamma - 1)}{2} M_e^2 f'\right](1 - f') - (1 - f')\right\} d\eta. \tag{46}$$

The above expressions are valid with heat transfer at the wall. For the case of no heat transfer at the wall one should use the expression

$$g_w = 1 + \frac{(\gamma - 1)}{2} M_e^2 \tag{47}$$

in equations (45) and (46).

From equations (42), (43) and (44) we may write

$$\frac{\Delta p_v}{p_e} = \gamma \left(\frac{1}{a_e} \frac{\partial \delta_Q}{\partial t} + M_e \frac{\partial \delta^*}{\partial x} \right). \tag{48}$$

On differentiation, expressions (45) and (46) give, respectively:

$$M_e \frac{\partial \delta^*}{\partial x} = \frac{1}{\sqrt{2}\, M_e^2} \bar{\chi} \left(I - 2\tau \frac{\partial I}{\partial \tau} \right) \tag{49}$$

and

$$\frac{1}{a_e} \frac{\partial \delta_Q}{\partial t} = \frac{\sqrt{2}}{M_e^2} \bar{\chi} \frac{\partial J}{\partial \tau} \tag{50}$$

where

$$I = \int_0^e \left[g_w + \frac{(\gamma - 1)}{2} M_e^2 f' \right] (1 - f')\, d\eta\,, \tag{51}$$

$$J = \int_0^e \left\{ \left[g_w + \frac{(\gamma - 1)}{2} M_e^2 f' \right] (1 - f') - (1 - f') \right\} d\eta \tag{52}$$

and

$$\bar{\chi} = M_e^3 \sqrt{\frac{\nu_e C}{U_e x}}\,. \tag{53}$$

Substitution of expressions (49) and (50) in (48) yields

$$\frac{\Delta p_v}{p_e} = \frac{\gamma}{M_e^2} \left[\sqrt{2} \frac{\partial J}{\partial \tau} + \frac{1}{\sqrt{2}} \left(I - 2\tau \frac{\partial I}{\partial \tau} \right) \right] \bar{\chi}\,. \tag{54}$$

For the details of expression (54), one should see ref. [1]. In ref. [5], parameter λ has been used in place of $\bar{\chi}$, where λ and $\bar{\chi}$ are related by

$$\lambda = a \bar{\chi}\,. \tag{55}$$

In the above expression a depends primarily on wall temperature for the steady state case. One may also express equation (54) in terms of the parameter λ as:

$$\frac{\Delta p_v}{p_e} = \lambda = a \bar{\chi}\,. \tag{56}$$

Here a depends on g_w and τ for given γ and M_e. Expression (56) is similar to expression (8) of reference [6] where the steady state induced pressure results have been obtained.

From expression (54) we may also write the expression for the transient contribution to the induced pressure referenced to the final steady state as

$$\frac{\Delta p}{p_e} = \frac{\Delta p_{v2} - \Delta p_v}{p_e} = \frac{\gamma}{M_e^2} \left[\frac{I_2}{\sqrt{2}} - \sqrt{2} \frac{\partial J}{\partial \tau} - \frac{1}{\sqrt{2}} \left(I - 2\tau \frac{\partial I}{\partial \tau} \right) \right] \bar{\chi}\,. \tag{57}$$

Numerical results for the transient contribution have been obtained on IBM-360/67 computer. Various results are displayed in Figs. 3 through 7.

IX. Wall Shear Stress and Heat Transfer

The wall shear stress may be obtained from

$$\tau_w = \left(\mu \frac{\partial u}{\partial y} \right) \bigg|_{y=0}\,. \tag{58}$$

Using the transformations of section III, one may also write

$$\tau_w = \alpha_1 f''(0, \tau) \tag{59}$$

where

$$\alpha_1 = \frac{\mu_w \varrho_w}{\varrho_e} \frac{U_e^{3/2}}{\sqrt{2 \nu_e C x}} \tag{60}$$

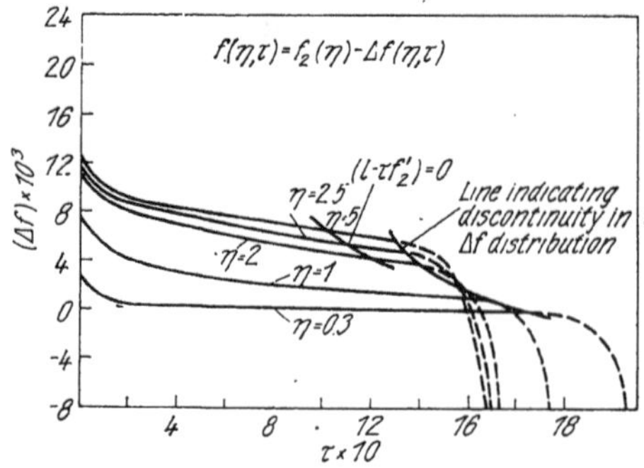

Fig. 1a. Stream Function Increment (Δf) Distribution Curve

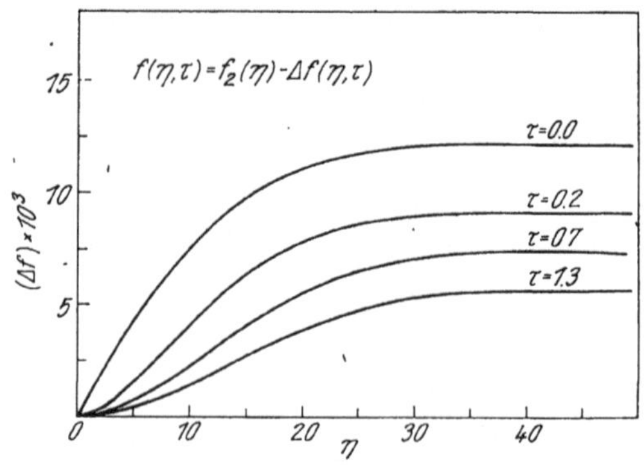

Fig. 1b. Profiles of Stream Function Increment Δf

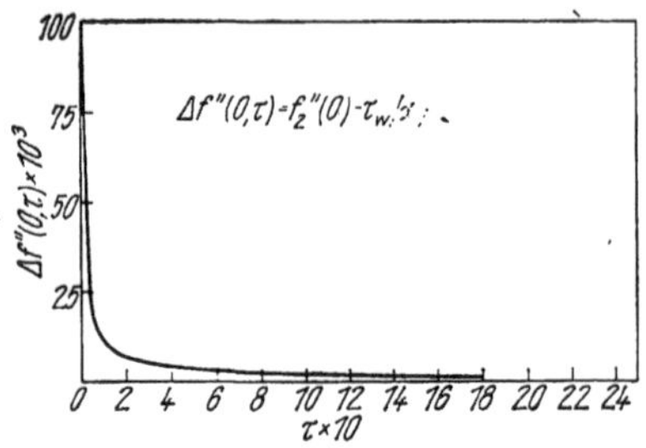

Fig. 2. Distribution of the Shearing Stress at the Wall

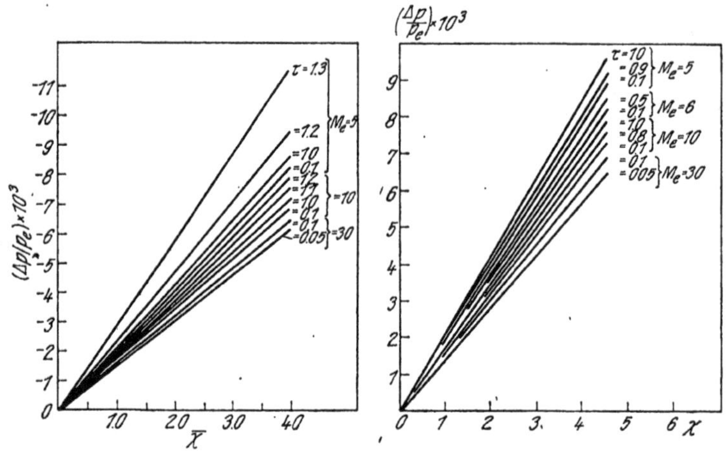

Fig. 3a/3b. Transient Contribution to the Induced Pressure for the Case of No Heat Transfer at the Wall
a) Stepwise-decelerated flow b) Stepwise-accelerated flow (Ref. 1)

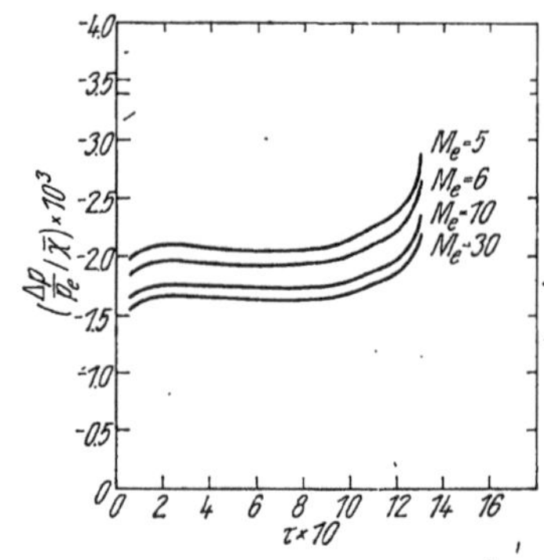

Fig. 3c. Transient Contribution to the Induced Pressure for the Case of No Heat Transfer

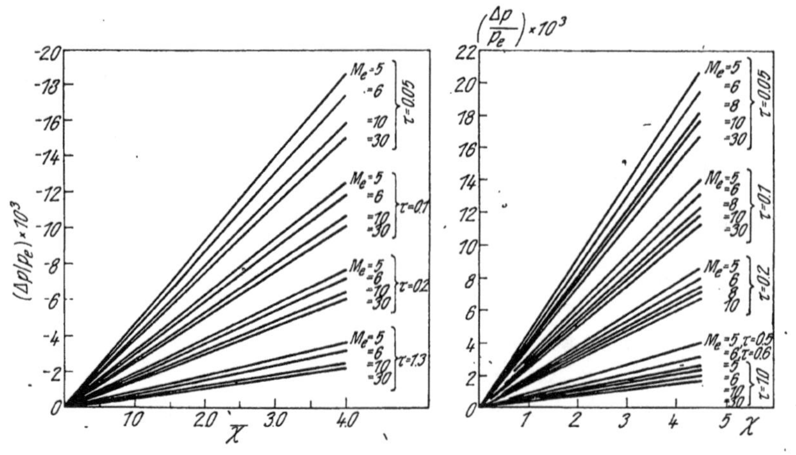

Fig. 4. Transient Contribution to the Induced Pressure for the Case of $g_w = 0.0$
a) Stepwise-decelerated flow b) Stepwise-accelerated flow (Ref. 1)

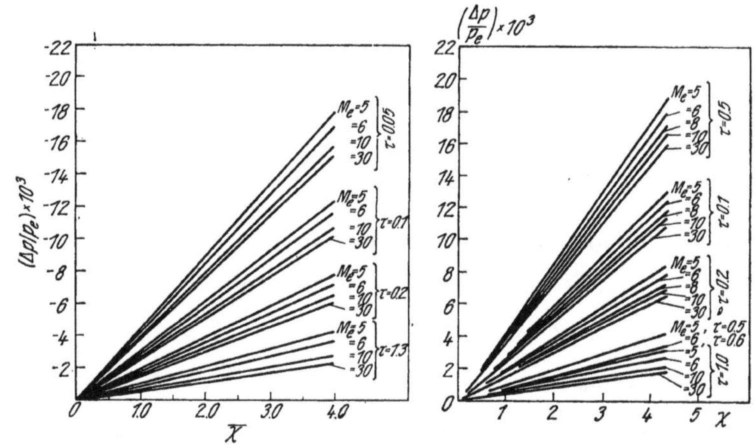

Fig. 5. Transient Contribution to the Induced Pressure for the Case of $g_w = 0.5$
a) Stepwise-decelerated flow b) Stepwise-accelerated flow (Ref. 1)

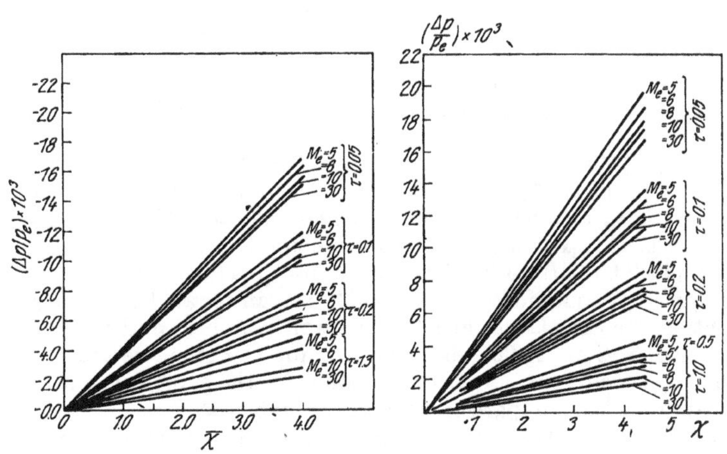

Fig. 6. Transient Contribution to the Induced Pressure for the Case of $g_w = 1.0$
a) Stepwise-decelerated flow b) Stepwise-accelerated flow (Ref 1)

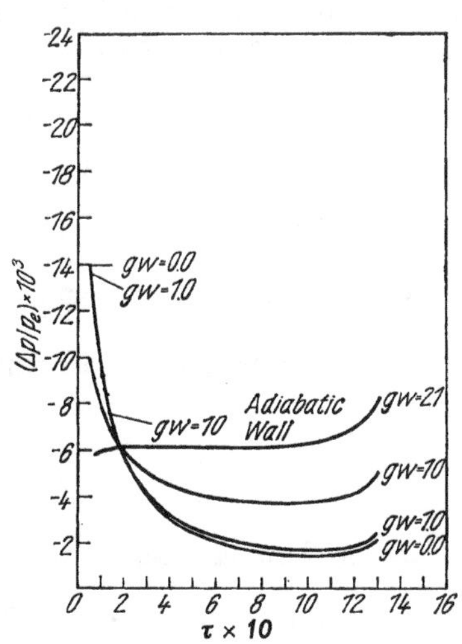

Fig. 7a. Transient Contribution to the Induced Pressure for the Case of $\bar{\chi} = 3.5$ and $M_e = 10$

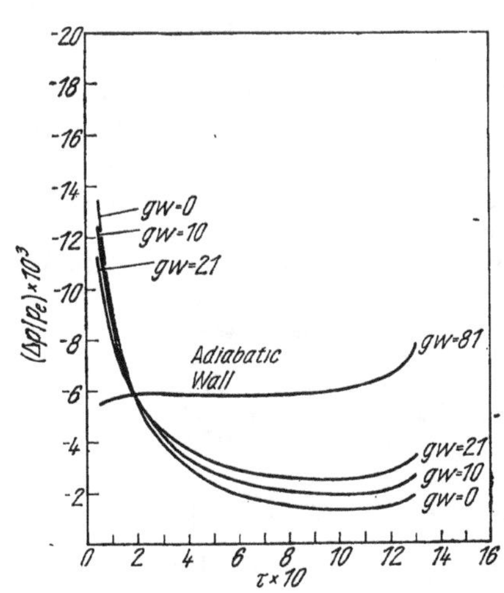

Fig. 7b. Transient Contribution to the Induced Pressure for the Case of $\bar{\chi} = 3.5$ and $M_e = 20$

Figure 2 shows a plot ot $\Delta f''(0,\tau)$ as a function of τ. One may easily obtain the wall shear stress from this plot. The heat transfer at the wall may be obtained from

$$g_w = \left(-k\frac{\partial T}{\partial y}\right)\Bigg|_{y=0} = \alpha_2 f''(0,\tau) \tag{61}$$

where

$$\alpha_2 = \frac{k_w h_e \varrho_w}{C_p \varrho_e}\sqrt{\frac{U_e}{2\nu_e C x}}\left[g_w - 1 - \frac{\gamma-1}{2}M_e^2\right]. \tag{62}$$

The wall heat transfer may also be calculated from Fig. 2 using the relation:

$$g_w = \frac{\alpha^2}{\alpha_1}\tau_w\,.$$

X. Discussion of the Results and Conclusions

The present work was undertaken primarily to investigate differences between the stepwise-accelerated and the stepwise-decelerated type of high Mach Number flows and to improve the existing numerical method used in the computations.

Following the approach of reference [1], the problem has been linearized. The stream function increment (Δf) contained in the linearized momentum equations has been obtained by taking one percent change in the free stream velocity. Double precision has been employed in the computations. This helped in obtaining the Δf-distribution up to $\tau = 1.3$ for $\eta = 5$. The other points of discontinuities are noticeable approximately at $\tau = 1.85$ for $\eta = 0.3$ At $\tau = 1.65$ for $\eta = 1$ and at $\tau = 1.4$ for $\eta = 2$.

The Δf-distribution for the stepwise-accelerated and the stepwise-decelerated flat plate cases has been found to agree with one another up to the 7th significant digit. Figures 1a and 1b show the Δf-distribution for the stepwise-decelerated flat-plate.

In Fig. 2 the transient distribution of $(\Delta f)''(0,\tau)$ is shown for $0 \leqq \tau \leqq 1.8$. This gives the desired shear stress distribution at the wall. For larger magnitudes of τ, $(\Delta f)''(0,\tau)$ appears to approach the final steady state value. From the same plot heat transfer at the wall may also be obtained as shown in section IX.

The comparison of the absolute values of this plot with the respective graph of ref. [1] indicates that the two cases differ only in the sixth significant digit. The $(\Delta f)''(0,\tau)$ distribution in ref. [1] is obtained for $0 \leqq \tau \leqq 1.3$.

The transient contribution to the induced pressure for the case of no heat transfer at the wall is given in Fig. 3a. The absolute magnitude of the contribution becomes larger at smaller Mach Numbers and bigger values of τ for a given $\bar{\chi}$. The contribution* in the range $5 \leqq M_e \leqq 30$, vis-a-vis that for the stepwise-accelerated flat plate problem of Ref. [1] is almost equal and larger for $0 \leqq \tau < 0.5$. It is smaller for the same range of Mach Numbers and $0.6 < \tau \leqq 1.3$. This may be observed from the comparison of Figs. 3a and 3b.

Figure 3c shows the transient contribution to the pressure distribution as a function of τ for the no heat transfer case.

For the case of non-zero wall heat transfer $\left(0 \leqq g_w < 1 + \left(\frac{\gamma-1}{2}\right)M^2\right)$, the transient contributions to the pressure distribution are given in Figs. 4a, 5a amd 6a. The absolute magnitude of these contributions is found to be larger for $0 \leqq \tau < 0.5$ and smaller for $0.6 < \tau \leqq 1.3$ as compared with the respective contributions of ref. [1]. These are given in Figs. 4b, 5b and 6b. It may be noted that the limits on τ for the cases of zero and non-zero wall heat transfer are identical.

In Figs. 7a and 7b some of the results for non-zero wall heat transfer have been further analysed for $\bar{\chi} = 3.5$, $M_e = 10$ and 20 respectively. It may be noticed that at a fixed Mach Number and fixed $\bar{\chi}$ the transient contribution is larger for lower values of g_w up to $\tau \cong 0.18$ whereas the transient contribution is smaller for lower values of g_w in the range $0.18 < \tau < 1.3$. For the case of an adiabatic wall the transient contribution to the pressure distribution is nearly constant up to $\tau = 1$.

For given values of g_w, $\bar{\chi}$ and τ, the transient contribution decreases with the increase in Mach Number. This may be noted again from Figs. 4a, 5a and 6a.

Finally, for the case of no heat transfer at the wall and $0 \leqq \tau \leqq 1.3$, it has been found in expression (57) that

$$\left|\frac{1}{\sqrt{2}}(I_2 - I)\right| \gg \left|\sqrt{2}\left(\frac{\partial J}{\partial \tau} - \tau\frac{\partial I}{\partial \tau}\right)\right|.$$

Also, for $0 \leqq g_w < 1$ and $\tau \ll 1$, it has been noticed that $\sqrt{2}\,\frac{\partial J}{\partial \tau}$ is the predominant term. All these findings are consistent with those of ref. [1] and, therefore, simplifications suggested therein for expression (57) are valid here also.

Acknowledgements

The financial assistance of the National Research Council of Canada through Grant No. NRC A-4198 is gratefully acknowledged.

*) It should be pointed out that all the comparison and other relevant statements are valid for absolute magnitudes only.

References

1 Rodkiewicz, C. M., and Reshotko, E., Time Dependent Hypersonic Viscous Interactions, US AFOSR 67-2451, 1967.
2 Stewartson, K., The Theory of Laminar Boundary Layers in Compressible Fluids, Oxford Mathematical Monographs, 1964.
3 Moore, F. K., Theory of Laminar Flows, High Speed Aerodynamics and Jet Propulsion, Vol. IV, Princeton University Press, pp. 429–524, 1964.
4 Moore, F. K., and Ostrach, S., Displacement Thickness of the Unsteady Boundary Layer, Journal of the Aero. Sci., January 1957.
5 White Jr., F. M., Hypersonic Laminar Viscous Interactions on Inclined Flat Plates, ARS Journal, pp. 780–781, May 1962.
6 Bertram, M. H. and Blackstock, T. A., Some Solutions to the Problem of Predicting Boundary-Layer Self-Induced Pressures, NASA TN D-798, April 1961.

Eingereicht am 16. 2. 1978

Anschriften: Cz. M. Rodkiewicz, Prof., Department of Mechanical Engineering, University of Alberta, Edmonton, Canada — R. N. Gupta, Prof., Department of Aeronautical Engineering, Indian Institute of Technology, Kanpur 208016, India

ZAMM 60, 587 –594 (1980)

DANG HONG TIEM

Ecoulement moléculaire libre autour d'un dièdre droit

Die freie Molekularströmung an einem konkaven rechtwinkligen Keil wird systematisch untersucht. Die ungestörte Gasströmung wird maxwellisch und die Reflektion an der Grenzfläche wird als vollkommen diffus vorausgesetzt. Es werden analytische Ausdrücke für die makroskopischen Größen der Gasströmung erhalten und die nummerischen Ergebnisse werden in Form von Linien gleicher Dichte, gleicher Temperatur und von Stromlinien wiedergegeben.

Free molecular flow over a concave rectangular wedge is systematically studied. The unperturbed gas flow is supposed maxwellian and the reflection on the boundary surfaces is supposed completely diffuse. Analytical expressions for macroscopic quantities of the gas flow are obtained and numerical results are given in forms of equal density lines, equal temperature lines, and current lines.

Исследуется свободное молекулярное течение около вогнутого препятствия, состоящего из двух перпендикулярных пластин. Невозмущённое течение газа предполагается максвелловым, а отражение на граничной поверхности — полностью диффузным. Получены аналитические выражения для макроскопических величин газового потока, а численные результаты представлены в виде линий одинаковой плотности, одинаковой температуры и линий тока.

1. Introduction

Les écoulements moléculaires libres sont devenus des sujets de recherche intense depuis une trentaine d'années. On a calculé surtout des forces aérodynamiques appliquées à un obstacle et le flux de chaleur transmis à celui-ci par les molécules gazeuses. On distingue en général deux cas suivant que l'obstacle est convexe ou concave. Dans le premier cas, presque toutes les formes géométriques simples de l'obstacle (plan, sphère, cylindre, etc. . . .) ont été déjà étudiées [1]. Dans le second cas, on a porté une attention particulière sur l'écoulement du gaz à travers des tubes de différentes formes (voir par exemple [2]).

Les corps concaves, autres que les tubes, ont été étudiés par M. T. CHAHINE [3] pour les cas des surfaces cylindriques et sphériques. Due à la forme particulière de ces surfaces, ainsi qu'à l'hypothèse que le flux de molécules incidentes est hypersonique (de vitesses uniformes), des résultats analytiques ont été obtenus pour la quantité de mouvement et pour l'énergie transmise à ces surfaces.

Le cas du corps concave constitué par deux plaques a été étudié par C. R. WIMBERLY [4], il a obtenu des résultats numériques pour les mêmes quantités précédentes.

W. WUEST [5] a étudié le cas d'un corps ayant la forme d'une fusée qui présente une partie concave, il a tracé des courbes d'égale densité autour de ce corps.

Dans ce qui suit, nous nous proposons d'étudier l'écoulement moléculaire libre autour d'un obstacle concave constituè par deux plaques perpendiculaires. C'est un problème que l'on rencontre souvent dans des études expérimentales. Des appareils de mesure ainsi que des corps volants, par exemple des fusées, présentent toujours des parties concaves plus ou moins de cette forme. On peut dire que, à l'exception de la résolution numérique du système de deux équations intégrales qui détermine les flux de molécules incidentes sur les deux plaques, tout le problème a été résolu analytiquement. Presque toutes les quantités physiques macroscopiques du gaz ont été calculées. Dans la résolution numérique des èquations intégrales de ce problème, qui sont singulières au voisinage de l'origine, une méthode particulière a été proposée pour obtenir des résultats avec une précision satisfaisante. Cette méthode ainsi que les résultats analytiques du problème peuvent s'étendre à un système de deux plaques quelconques.

2. Hypothèses et équations

On considère un flux gazeux de vitesse globale $\boldsymbol{U}_\infty$ et de température T_∞ à l'infini, qui tombe sur deux plaques immobiles et perpendiculaires, de température constante et uniforme T_w. On suppose que ce flux gazeux admet à l'infini une fonction de distribution maxwellienne, et que la réflexion sur les plaques est complètement diffuse, par conséquent les molécules réfléchies sont en équilibre avec les plaques et ont pour fonction de distribution une fonction maxwellienne. On suppose encore que le nombre de KNUDSEN $Kn = \lambda/a$, où λ est le libre parcours moyen des molécules, et a la largeur commune de deux plaques, est infini, de telle sorte que les effets dus aux collisions entre les molécules sont négligeables.

On se propose d'étudier l'écoulement gazeux autour de ces deux plaques. Le problème consiste à trouver la fonction de distribution du gaz en un point quelconque de l'espace.

La fonction de distribution du gaz à l'infini est de la forme

$$f_\infty = n_\infty \left(\frac{h_\infty}{\pi}\right)^{3/2} \exp\left[-h_\infty(\boldsymbol{\xi} - \boldsymbol{U}_\infty)^2\right] \tag{1}$$

où n_∞ est la densité du gaz et $h_\infty = m/2kT_\infty$, m étant la masse d'une molécule gazeuse et k la constante de BOLTZMANN.

La fonction de distribution des molécules réfléchies peut s'écrire sous une forme analogue

$$f_w = n_w \left(\frac{h_w}{\pi}\right)^{3/2} \exp\left(-h_w \xi^2\right) \tag{2}$$

où $h_w = m/2kT_w$ et $n_w/2$ la densité de ces molécules.

On se donne n_∞, T_∞, $\boldsymbol{U}_\infty$ et T_w, si on connait n_w, alors le problème est résolu. En effet, considérons les molécules qui arrivent en un point M quelconque, on divise l'espace en trois régions définies par les trois dièdres de sections droites α_0, α_1, α_2. Comme il n'y a pas de collisions entre les molécules, celles qui ont des vitesses ξ situées dans le dièdre de section droite α_0 arrivent de l'infini, et celles qui ont des ξ dans (α_1) sont réfléchies par la plaque Π_1, et soit n_{w1} la valeur de n_w relative à ces molécules, de même pour le dièdre (α_2) et la plaque Π_2, et soit n_{w2} la valeur correspondante de n_w (fig. 1), d'où la solution

$$\xi \in (\alpha_0) \quad f = f_\infty\,, \qquad \xi \in (\alpha_1) \quad f = f_{w1}\,, \qquad \xi \in (\alpha_2) \quad f = f_{w2}\,. \tag{3}$$

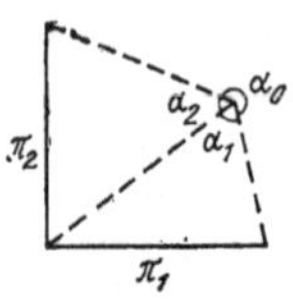

Fig. 1

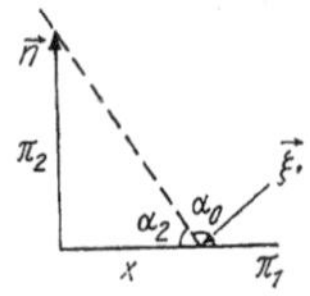

Fig. 2

Pour déterminer n_{w1} et n_{w2}, on utilise la condition qui exprime le fait qu'il n'y a pas d'accumulation des molécules sur les plaques, c'est-à-dire que le flux de molécules incidentes par unité de surface et de temps est égal à celui de molécules réfléchies. En un point de Π_1 par exemple, cette condition s'écrit

$$\int_{\xi\cdot\boldsymbol{n}>0} (\xi\cdot\boldsymbol{n})\, f_{w1}\, \mathrm{d}\xi = -\int_{\xi\in(\alpha_0)} (\xi\cdot\boldsymbol{n})\, f_\infty\, \mathrm{d}\xi - \int_{\xi\in(\alpha_2)} (\xi\cdot\boldsymbol{n})\, f_{w2}\, \mathrm{d}\xi \tag{4}$$

où (α_0) et (α_2) désignent des dièdres de sections droites α_0 et α_2 (fig. 2), et $\boldsymbol{n}$ est le vecteur unitaire normal à Π_1 et dirrigé vers le gaz.

Dans la pratique, au lieu de n_{w1} par exemple, on préfère définir la quantité N_1 qui est le nombre de molécules incidentes sur la plaque Π_1 par unité de surface et de temps, on a

$$N_1 = \int_{\xi\cdot n>0} (\xi\cdot\boldsymbol{n})\, f_{w1}\, \mathrm{d}\xi = \frac{n_{w1}}{2\sqrt{\pi h_{w1}}} \tag{5}$$

et la condition de la non-accumulation des molécules gazeuses sur les plaques nous donne deux équations intégrales suivantes [6]

$$\begin{aligned} N_1(x) &= N_{1\infty}(x) + \frac{1}{2}\int_0^1 \frac{xyN_2(y)}{(x^2+y^2)^{3/2}}\,\mathrm{d}y\,, \\ N_2(y) &= N_{2\infty}(y) + \frac{1}{2}\int_0^1 \frac{xyN_1(y)}{(x^2+y^2)^{3/2}}\,\mathrm{d}x \end{aligned} \tag{6}$$

où x et y sont des abscisses des points de Π_1 et Π_2 respectivement, la largeur commune de Π_1 et Π_2 est prise égale à 1. $N_{1\infty}(x)$ est le nombre de molécules incidentes sur Π_1 venant de l'infini, il est donné par l'expression

$$N_{1\infty}(x) = -\int_{\xi\in(\alpha_0)} (\xi\cdot\boldsymbol{n})\, f_\infty\, \mathrm{d}\xi = \frac{n_\infty}{4\sqrt{\pi h_\infty}}\{S\sin\alpha[I(\alpha) + I(\beta)] + E(\alpha) - \cos\alpha_0 E(\beta)\} \tag{7}$$

où $S = \sqrt{h_\infty}\, U_\infty$, α est l'angle de $-\boldsymbol{U}$ avec Π_1, α_0 est défini par $\tan\alpha_0 = -1/x$, $\beta = \alpha_0 - \alpha$,

$$\begin{aligned} I(u) &= \frac{2u}{\sqrt{\pi}} + \sqrt{\pi}\,\mathrm{erf}\,(S\sin u) + 2\int_0^{S\sin u} \mathrm{e}^{-t^2}\,\mathrm{erf}\,(t\cot u)\,\mathrm{d}t\,, \\ E(u) &= \exp\left(-S^2\sin^2 u\right)[1 + \mathrm{erf}\,(S\cos u)]\,. \end{aligned} \tag{8}$$

Le second terme représente le nombre de molécules incidentes à Π_1 (resp. Π_2) par réflexion sur Π_2 (resp. Π_1)

$$\frac{1}{2}\int_0^1 \frac{xyN_2(y)}{(x^2+y^2)^{3/2}}\,\mathrm{d}y = -\int_{\xi\in(\alpha_2)} (\xi\cdot\boldsymbol{n})\, f_{w2}\,\mathrm{d}\xi$$

et enfin, $N_{2\infty}(y)$ est obtenu de $N_{1\infty}(x)$ en remplaçant α par $\pi/2 - \alpha$.

3.1. Cas particulier

Le système (6) admet une solution simple dans le cas où le gaz est au repos à l'infini et dans le cas où la largeur de deux plaques peut être considérée comme infinie.

Dans le premier cas, $U_\infty = 0$,

$$N_{1\infty}(x) = \frac{n_\infty}{4\sqrt{\pi h_\infty}}(1 - \cos\alpha_0) = \frac{n_\infty}{4\sqrt{\pi h_\infty}}\left(1 + \frac{x}{\sqrt{1 + x^2}}\right) \tag{9}$$

le système (6) admet la solution

$$N_1 = N_2 = \frac{n_\infty}{2\sqrt{\pi h_\infty}}. \tag{10}$$

On peut en déduire des grandeurs macroscopiques du gaz en un point quelconque de l'espace, en particulier (fig. 1)

$$n = \frac{n_\infty}{2\pi}\left(\alpha_0 + \frac{\alpha_1 + \alpha_2}{\sqrt{\tau}}\right), \qquad T = T_\infty \frac{\alpha_0 + \sqrt{\tau}(\alpha_1 + \alpha_2)}{\alpha_0 + (\alpha_1 + \alpha_2)/\sqrt{\tau}}, \qquad u = 0, \tag{11}$$

où on a posé $\tau = T_w/T_\infty$.

Dans le cas où la largeur de deux plaques est infinie, $\alpha_0 = \pi/2$, on a

$$N_{1\infty} = \frac{n_\infty}{4\sqrt{\pi h_\infty}}\{1 + \operatorname{erf}(S\cos\alpha)\}\{\exp(-S^2\sin^2\alpha) + \sqrt{\pi}\, S\sin\alpha\,[1 + \operatorname{erf}(S\sin\alpha)]\} \tag{12}$$

qui est une constante, le système (6) où on a remplacé la borne 1 des intégrales par ∞, admet la solution

$$N_1 = \tfrac{2}{3}(2N_{1\infty} + N_{2\infty}), \qquad N_2 = \tfrac{2}{3}(N_{1\infty} + 2N_{2\infty}). \tag{13}$$

3.2. Cas général

Nous allons étudier le système (6) dans le cas général, il s'écrit

$$N_1(x) = N_{1\infty}(x) + \tfrac{1}{2}\int_0^1 K(x, y)\, N_2(y)\, dy, \qquad N_2(y) = N_{2\infty}(y) + \tfrac{1}{2}\int_0^1 K(x, y)\, N_1(x)\, dx \tag{14}$$

avec

$$K(x, y) = \frac{xy}{(x^2 + y^2)^{3/2}}.$$

Notons que le noyau $K(x, y)$, non borné, ne vérifie pas la condition

$$\int_0^1\int_0^1 |K(x, y)|^2\, dx\, dy < +\infty$$

qui est en général admise dans la théorie des équations intégrales, c'est pourquoi il nous faut établir l'existence et l'unicité de la solution directement à partir de ce système. Pour cela, il faut trouver quelques propriétés analytiques des fonctions connues $N_{1\infty}(x)$ et $N_{2\infty}(y)$.

a) Properiétés de $N_{1\infty}(x)$.

On calcule d'abord les dérivées première et seconde de $N_{1\infty}(x)$ par rapport à x, les calculs sont très compliqués et ne sont pas donnés dans ce travail, on écrit ici les résultats

$$N'_{1\infty} = \frac{n_\infty}{4\sqrt{\pi h_\infty}}\sin^3\alpha_0\left[\frac{2S}{\sqrt{\pi}}e^{-S^2}\cos\beta + (1 + 2S^2\cos^2\beta)\,E(\beta)\right],$$

$$N''_{1\infty} = \frac{n_\infty}{4\sqrt{\pi h_\infty}}\sin^4\alpha_0\left\{3\cos\alpha_0\left[\frac{2S}{\sqrt{\pi}}e^{-S^2}\cos\beta + (1 + 2S^2\cos^2\beta)\,E(\beta)\right] - \right.$$
$$\left. - 2S\sin\alpha_0\sin\beta\left[\frac{2}{\sqrt{\pi}}e^{-S^2}(1 + S^2\cos^2\beta) + S\cos\beta\,(3 + 2S^2\cos^2\beta)\,E(\beta)\right]\right\}.$$

On peut démontrer d'une facon rigoureuse que $N'_{1\infty}$ est toujours positive et que $N''_{1\infty}$ est toujours négative, par conséquent la courbe représentant la variation de $N_{1\infty}$ en fonction de x est croissante et tourne sa concavité vers le bas. On en déduit les inégalités suivantes

$$N_{1\infty}(x) \leqq N_{1\infty}(1), \qquad N_{1\infty}(x) \leqq N_{1\infty}(0) + xN'_{1\infty}(0). \tag{15}$$

b) Unicité de la solution.

Nous nous proposons de démontrer que si le système (14) admet une solution bornée, alors cette solution est unique.

En effet, soient (N_1, N_2) et (N_1^*, N_2^*) deux solutions de ce système, on a

$$N_1 - N_1^* = \tfrac{1}{2}\int_0^1 K(x, y)\,(N_2 - N_2^*)\, dy. \tag{16}$$

Soit M la plus grande de deux bornes supérieures de $(N_1 - N_1^*)$ et $(N_2 - N_2^*)$, on a

$$|N_1 - N_1^*| \leqq \frac{1}{2}\int_0^1 K(x, y)\,|N_2 - N_2^*|\, dy \leqq \frac{M}{2}\int_0^1 K(x, y)\, dy \leqq \frac{M}{2} \tag{17}$$

de même

$$|N_2 - N_2^*| \leqq \frac{M}{2}.$$

En répétant les calculs avec la borne $M/2$, etc. . . ., on trouve, après n calculs

$$|N_1 - N_1^*| \leqq M/2^n, \qquad |N_2 - N_2^*| \leqq M/2^n$$

et ceci est vrai quel que soit n, d'où

$$N_1 = N_1^*, \qquad N_2 = N_2^*.$$

c) Existence de la solution.

On cherche la solution du système par la méthode d'itération a

$$N_1(x) = N_{1\infty}(x) + \frac{1}{2}\int_0^1 K(x, y)\, N_{2\infty}(y)\, dy + \frac{1}{2^2}\int_0^1\int_0^1 K(x, y)\, K(y, u)\, N_{1\infty}(u)\, dy\, du + \dots \tag{18}$$

On a établi que

$$N_{1\infty}(x) \leqq N_{1\infty}(1), \qquad N_{2\infty}(y) \leqq N_{2\infty}(1)$$

soit

$$A = \sup\{N_{1\infty}(1), N_{2\infty}(1)\}$$

la série (18) est majorée par

$$A + \frac{A}{2}\int_0^1 K(x, y)\, dy + \frac{A}{2^2}\int_0^1\int_0^1 K(x, y)\, K(y, u)\, dy\, du + \dots$$

qui est à son tour majorée par

$$A + \frac{A}{2}\int_0^\infty K(x, y)\, dy + \frac{A}{2^2}\int_0^\infty\int_0^\infty K(x, y)\, K(y, u)\, dy\, du + \dots \tag{19}$$

les intégrales précédentes peuvent être calculées effectivement, elles sont toutes égales à 1, d'où cette série s'écrit

$$A + \frac{A}{2} + \frac{A}{2^2} + \dots = 2A.$$

Donc la série (18) converge uniformément. Il nous reste à démontrer que (18) vérifie effectivement le système (14), il suffit pour cela de porter la série (18) dans les deux membres de (14) et vérifier que ces deux membres sont identiques.

d) Limites de $N_1(x)$ et $N_2(y)$ pour x et y tendant vers 0.

On ne peut trouver la solution du système sous forme analytique condensée, mais on peut déterminer la limite de la solution au voisinage de l'origine, cette limite servira de point de repère pour des résultats numériques.

Nous allons démontrer que pour $x \to 0$, et $y \to 0$

$$N_1(x) \to \tfrac{2}{3}\,[2N_{1\infty}(0) + N_{2\infty}(0)], \qquad N_2(y) \to \tfrac{2}{3}\,[N_{1\infty}(0) + 2N_{2\infty}(0)] \tag{20}$$

c'est-à-dire quand on se trouve au voisinage de l'origine, les deux plaques peuvent être considérées comme infinies. Bien que ce résultat soit évident au point de vue physique, il est difficile de le démontrer mathématiquement, car dans l'équation

$$N_1(x) = N_{1\infty}(x) + \tfrac{1}{2}\int_0^1 K(x, y)\, N_2(y)\, dy$$

par exemple, on ne peut pas remplacer $N_2(y)$ par des approximations (y n'est pas petit, il varie de 0 à 1).

En faisant le changement des fonctions inconnues

$$N_1(x) = \tfrac{2}{3}\,[2N_{1\infty}(0) + N_{2\infty}(0)] + F_1(x), \qquad N_2(y) = \tfrac{2}{3}\,[N_{1\infty}(0) + 2N_{2\infty}(0)] + F_2(y) \tag{21}$$

le système devient

$$F_1(x) = A_1(x) + \tfrac{1}{2}\int_0^1 K(x, y)\, F_2(y)\, dy, \qquad F_2(y) = A_2(y) + \tfrac{1}{2}\int_0^1 K(x, y)\, F_1(x)\, dx \tag{22}$$

où $A_1(x)$ par exemple est égal à

$$A_1(x) = N_{1\infty}(x) - N_{1\infty}(0) - \frac{1}{3}\,[N_{1\infty}(0) + 2N_{2\infty}(0)]\,\frac{x}{(1 + x^2)^{1/2}} \tag{23}$$

et d'après (15), $A_1(x)$ peut être majoré par

$$|A_1(x)| < a_1 x \quad \text{avec} \quad a_1 = N'_{1\infty}(0) + \tfrac{1}{3}\,[N_{1\infty}(0) + 2N_{2\infty}(0)] \tag{24}$$

mais cette majoration doit être légèrement modifiée pour pouvoir être utilisée dans la suite. Soit $0 < \lambda < 1$, comme $0 \leqq x \leqq 1$, on peut écrire

$$|A_1(x)| \leqq a_1 x^\lambda \tag{25}$$

de même

$$|A_2(x)| \leqq a_2 x^\lambda \,.\,.$$

Nous nous proposons de démontrer que $F_1(x)$ tend vers 0 avec x. Par la méthode d'itération, on a

$$F_1(x) = A_1(x) + \frac{1}{2}\int_0^1 K(x, y)\, A_2(y)\, dy + \frac{1}{2^2}\int_0^1\int_0^1 K(x, y)\, K(y, u)\, A_1(u)\, dy\, du + \dots \tag{26}$$

d'où on peut majorer $F_1(x)$ par

$$|F_1(x)| \leqq a x^\lambda + \frac{1}{2}\, a \int_0^1 K(x, y)\, y^\lambda\, dy + \frac{1}{2^2}\, a \int_0^1\int_0^1 K(x, y)\, K(y, u)\, u^\lambda\, dy\, du + \dots \tag{27}$$

avec $a = \sup \{a_1, a_2\}$.

On désigne par $G_1(x)$ la première intégrale de (27), $G_2(x)$ la seconde intégrale, etc. . . . Posons $y = xt$,

$$G_1(x) = \int_0^1 K(x, y)\, y^\lambda\, dy = \int_0^{1/x} \frac{x^\lambda t^{1+\lambda}\, dt}{(1 + t^2)^{3/2}} < \int_0^\infty \frac{x^\lambda t^{1+\lambda}\, dt}{(1 + t^2)^{3/2}} = k x^\lambda \tag{28}$$

avec

$$k = \int_0^\infty \frac{t^{1+\lambda}\, dt}{(1 + t^2)^{3/2}} = \frac{\lambda}{2\sqrt{\pi}}\, \Gamma\left(\frac{\lambda}{2}\right) \Gamma\left(\frac{1}{2} - \frac{\lambda}{2}\right), \tag{29}$$

k est fonction de λ. Dans la suite, on doit restreindre à k la condition $k < 2$. On a

$$\frac{dk}{d\lambda} = \int_0^\infty \frac{t^{1+\lambda} \log t}{(1 + t^2)^{3/2}}\, dt = \int_0^1 (t^{2\lambda+1} - 1)\, \frac{\log t\, dt}{t^\lambda (1 + t^2)^{3/2}}\,.$$

On en déduit $dk/d\lambda > 0$, donc k est une fonction croissante de λ de $k(0) = 1$ à $k(1) = \infty$, d'où il existe une valeur λ_0 ($= 0{,}536$) telle que $k(\lambda_0) = 2$, et on doit choisir $\lambda < \lambda_0$.

De même, pour la seconde intégrale,

$$G_2(x) = \int_0^1 K(x, y)\, K(y, u)\, u^\lambda\, dy\, du < \int_0^1 k y^\lambda\, K(x, y)\, dy < k^2 x^\lambda \tag{30}$$

etc. . . ., d'où

$$|F_1(x)| < a x^\lambda \left(1 + \frac{k}{2} + \frac{k^2}{2^2} + \dots\right) = \frac{2 a x^\lambda}{2 - k} \tag{31}$$

donc $F_1(x)$ tend vers 0 avec x.

e) Application de la méthode de WIENER-HOPF.

En faisant le changement de variables $x = e^{-\lambda/2}$ et $y = e^{-\mu/2}$, et en posant

$$N_{1,1\infty}(x) = N^*_{1,1\infty}(\lambda)\,, \qquad N_{2,2\infty}(y) = N^*_{2,2\infty}(\mu)$$

le système (14) devient

$$N^*_1(\lambda) = N^*_{1\infty}(\lambda) + \tfrac{1}{4}\int_0^\infty k(\lambda - \mu)\, N^*_2(\mu)\, d\mu\,, \qquad N^*_2(\mu) = N^*_2(\mu) + \tfrac{1}{4}\int_0^\infty k(\mu - \lambda)\, N^*_1(\lambda)\, d\lambda \tag{32}$$

où

$$k(x) = \frac{e^x}{(1 + e^x)^{3/2}}\,.$$

La transformée de FOURIER de $k(x)$ peut être calculée facilement,

$$k^*(s) = \frac{1}{\sqrt{2\pi}} \int_{-\infty}^\infty \frac{e^{x - isx}}{(1 + e^x)^{3/2}}\, dx = \frac{1}{\sqrt{2\pi}} \int_0^\infty \frac{t^{-is}}{(1 + t)^{3/2}}\, dt = \frac{\sqrt{2}}{\pi}\, \Gamma(1 - is)\, \Gamma\left(\frac{1}{2} + is\right). \tag{33}$$

L'intégrale est convergente pour $-1 < \operatorname{Im} s < \frac{1}{2}$, par conséquent quand on fait tendre $|s|$ vers ∞, on peut négliger $\operatorname{Im} s$ devant $\operatorname{Re} s$, c'est-à-dire en supposant que s est réel, dans ce cas, on peut calculer le module de $k^*(s)$,

$$|k^*(s)|^2 = \frac{4s}{\sinh 2\pi s}$$

on trouve que $k^*(s)$ tend vers 0 avec $1/s$, d'où on peut appliquer la méthode de projection de WIENER-HOPF, mais je ne peux pas pousser les calculs analytiques plus loin.

4. Grandeurs macroscopiques et résultats numériques

Si on suppose connus $N_1(x)$ et $N_2(y)$, on peut calculer des grandeurs macroscopiques du flux gazeux en un point quelconque de l'espace. Dans ce qui suit, on donne seulement quelques expressions de ces grandeurs. Pour simplifier les formules à écrire, on pose

$$N = \frac{n_\infty}{4\sqrt{\pi h_\infty}}\,\overline{N}\,, \qquad n = \frac{n_\infty}{4\sqrt{\pi}}\,\overline{n}\,, \qquad u = \frac{\overline{u}}{\sqrt{h_\infty}}\,, \qquad T = \frac{2}{3}\,T_\infty \overline{T}\,.$$

La densité du gaz est égale à

$$\overline{n} = I(\alpha) + I(\beta) + \frac{1}{\sqrt{\pi\tau}}\int_0^1 \frac{y\overline{N}_1(t)\,\mathrm{d}t}{(x-t)^2+y^2} + \frac{1}{\sqrt{\pi\tau}}\int_0^1 \frac{x\overline{N}_2(t)\,\mathrm{d}t}{x^2+(y-t)^2}. \tag{34}$$

Les lignes de courant sont définies par l'équation

$$\psi = \text{constante}$$

avec

$$\psi = \sqrt{(x-1)^2+y^2}\,[S\sin\gamma_1 I(\gamma_1) + E(\gamma_1)] - \sqrt{x^2+(y-1)^2}\,[S\sin\gamma_2 I(\gamma_2) + E(\gamma_2)] + \\ + \frac{1}{2}\int_0^1 \frac{(x-t)\,\overline{N}_1(t)\,\mathrm{d}t}{\sqrt{(x-t)^2+y^2}} + \frac{1}{2}\int_0^1 \frac{(y-t)\,\overline{N}_2(t)\,\mathrm{d}t}{\sqrt{x^2+(y-t)^2}} \tag{35}$$

dans cette expression, les variables γ_1 et γ_2 sont définies comme dans la fig. 3. La vitesse globale du gaz est donnée par

$$\overline{n}\overline{u}_x = -\frac{\partial\psi}{\partial y}, \qquad \overline{n}\overline{u}_y = \frac{\partial\psi}{\partial x}, \qquad \overline{u}_z = 0\,. \tag{36}$$

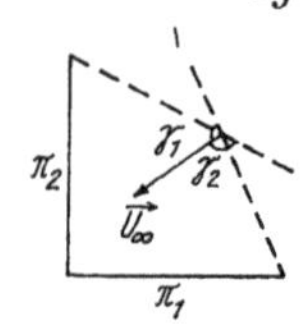

Fig. 3

La température du gaz est égale à

$$\overline{n}\overline{T} = \left(\frac{3}{2} + S^2\right)[I(\gamma_1) + I(\gamma_2)] + S\sin\gamma_1 E(\gamma_1) + S\sin\gamma_2 E(\gamma_2) - \overline{n}\,\overline{u}^2 + \\ + \frac{3}{2}\sqrt{\frac{\tau}{\pi}}\int_0^1 \left[\frac{y\overline{N}_1(t)}{(x-t)^2+y^2} + \frac{x\overline{N}_2(t)}{x^2+(y-t)^2}\right]\mathrm{d}t\,.$$

Enfin, on peut reprendre le problème avec les mêmes hypothèses sur le flux gazeux et sur la réflexion sur les plaques que précédemment, mais avec une ouverture au sommet de deux plaques, les calculs sont analogues quoique plus compliqués.

Pour la résolution numérique du système (14), on a rencontré une difficulté dans le calcul de l'intégrale au second membre, ceci résulte de la remarque que le calcul approché de l'intégrale

$$J = \int_0^1 \frac{xy\,\mathrm{d}y}{(x^2+y^2)^{3/2}} \tag{37}$$

pour la valeur de x égale à l'abscisse du point divisionnaire le plus proche de l'origine, donne un résultat avec une erreur toujours égale à 1/100, quelque soit le nombre de divisions de l'intervalle (0, 1). C'est par l'intègration par partie

$$J = \frac{1}{x\sqrt{1+x^2}} - \frac{1}{x}\int_0^1 \frac{y\,\mathrm{d}y}{\sqrt{x^2+y^2}} \tag{38}$$

que l'on peut réduire cette erreur à 10^{-4}.

Donc au lieu du système (14), on se ramène à résoudre le système suivant

$$N_1(x) = N_{1\infty}(x) + \frac{N_2(1)}{2x\sqrt{1+x^2}} - \frac{1}{2x}\int_0^1 \frac{[N_2(y) + yN_2'(y)]\,y}{\sqrt{x^2+y^2}}\,\mathrm{d}y \tag{39}$$

et une équation analogue pour $N_2(y)$.

Ce système se résoud par la méthode d'itération, à partir de $N_{1,2}^{0'} = N_{1\infty,2\infty}$. Les dérivées qui se trouvent au second membre sont calculées par la formule d'extrapolation polynomiale.

Dans la suite, nous donnons des résultats numériques représentés sous formes des courbes d'égale densité, d'égale température, et des lignes de courant.

Fig. 4: Courbe d'égale densité, $S = 1$, $\alpha = 45°$, $T_w/T_\infty = 1/10$. La valeur de la densité du gaz à l'infini est égale à $n_\infty = 4\sqrt{\pi} = 7{,}09$.

Figs. 5—7: Courbes d'égale température. $T_\infty = 1{,}5$ et $S = 1$. Les deux premières figures correspondent à $\alpha = 45°$, $T_w/T_\infty = 10$ et 1 respectivement. La fig. 7 correspond à $\alpha = 0°$ et $T_w/T_\infty = 1$. On constate que dans les cas où le gaz à l'infini et les deux plaques ont même température, il y a un maximum de température dans le flux gazeux.

Figs. 8—10: Lignes de courant, $S = 1$, $\alpha = 0°$, $30°$, $45°$. Dans le dernier cas, il y a un point d'arrêt, la ligne de courant passant par ce point divise le plan en deux régions, la région extérieure où toutes les lignes de courant, issues de l'infini, sont déviées vers l'extérieur, et la région intérieure où se trouvent deux familles de lignes de circulation fermées. L'ordre de grandeur moyen de la vitesse globale du gaz dans la région intérieure est égal à 1/100 de celui de la vitesse dans la région extérieure. Dans les trois cas, près de la plaque, les lignes de courant se concentrent dans une couche très mince dont l'épaisseur est certainement inférieure à 10^{-4}.

Figs. 11—12: Lignes de courant dans le cas où il existe une ouverture au sommet de deux plaques, de largeur égale à 0,2. $S = 1$ et $\alpha = 0°$ et $45°$.

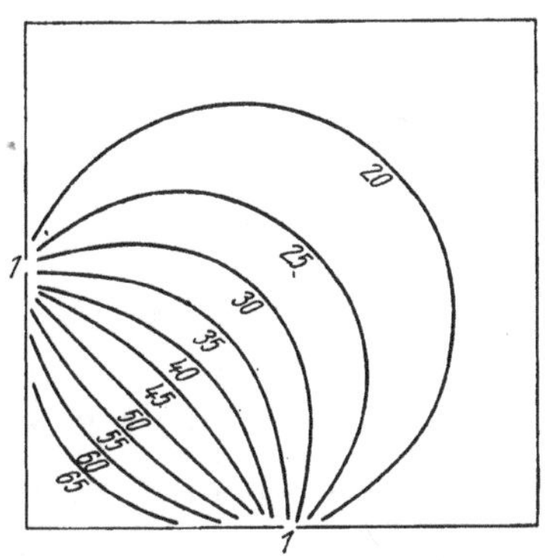

Fig. 4

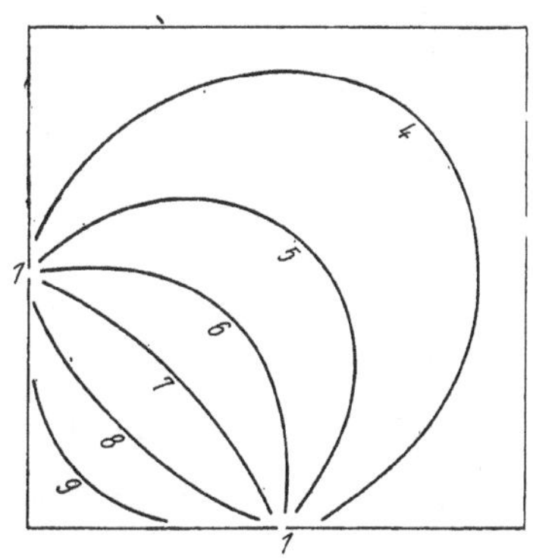

Fig. 5

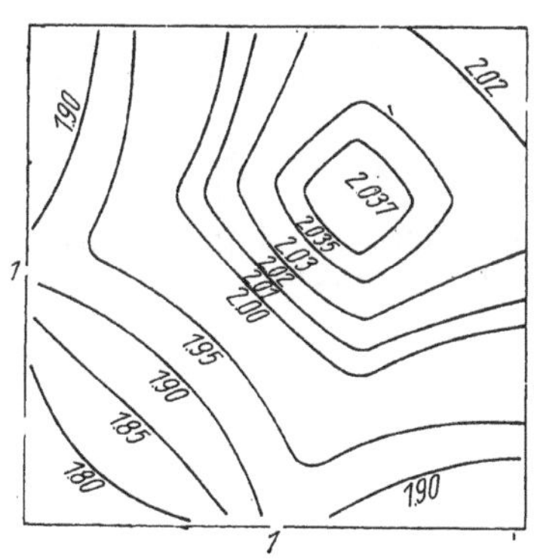

Fig. 6

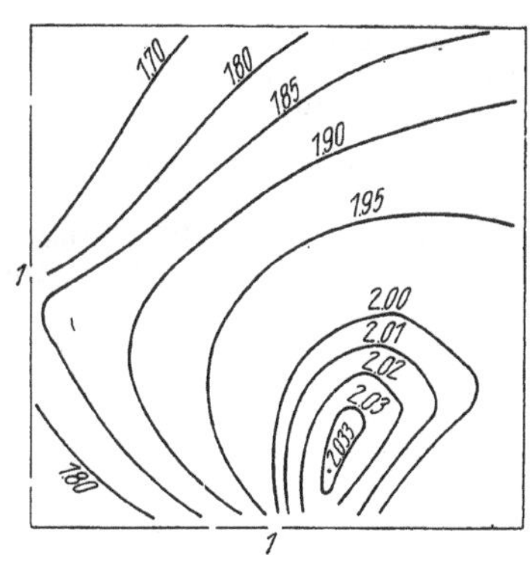

Fig. 7

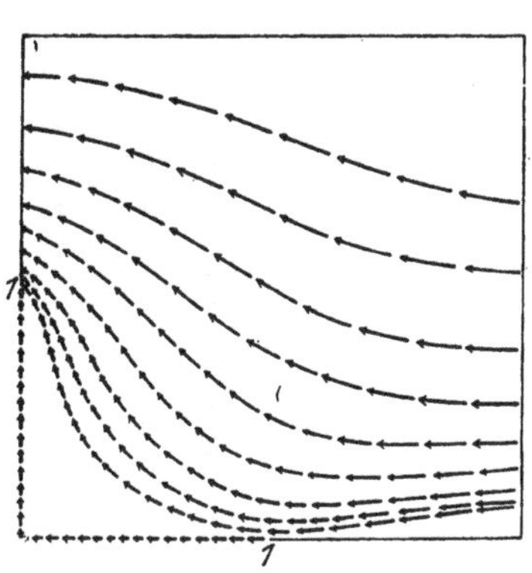

Fig. 8

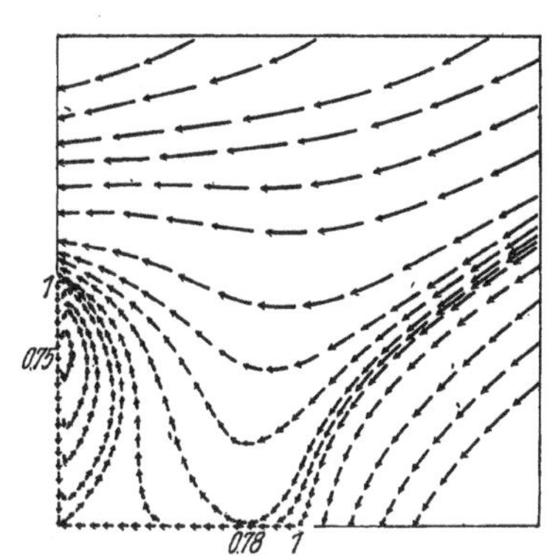

Fig. 9

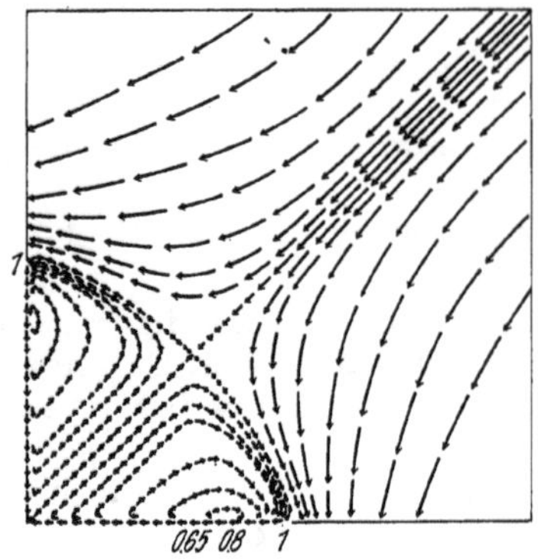

Fig. 10

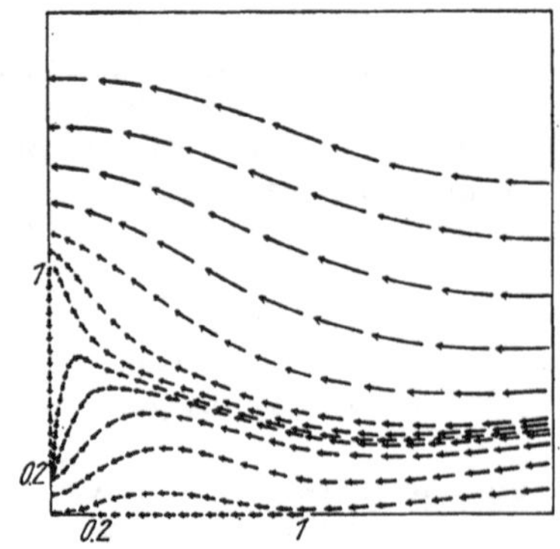

Fig. 11

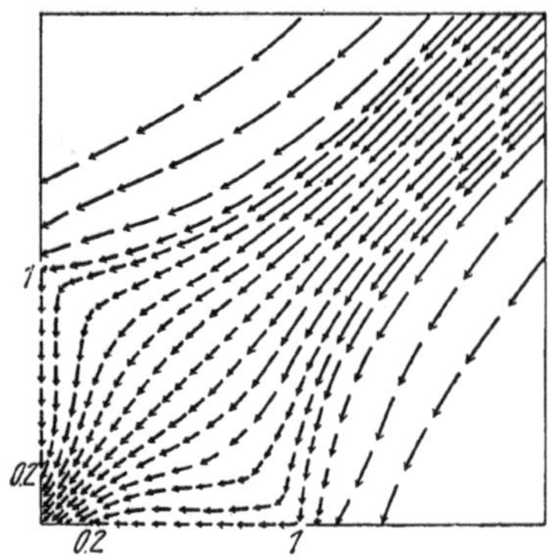

Fig 12

Bibliographies

1 SCHAAF, S. A., CHAMBE, P. L., Flow of Rarefied Gases, in Fundamentals of Gas Dynamics, edited by H. W. EMMONS, Princeton University Press, 1958.
2 EDWARDS, R. H., Low Density Flows Through Tubes and Nozzles, in 10th Symposium on Rarefied Gas Dynamics, edited by J. L. POTTER, 1977, p. 199.
3 CHAHINE, M. T., Free Molecule Flow over Nonconvexe Surface, in 2nd Symposium on Rarefied Gas Dynamics, edited by L. TALBOT, Academic Press, 1961, p. 209.
4 WIMBERLY, C. R., Effect of Molecular Interreflections on a Concave Wedge in Free Molecule Flow, AIAA Journal, 6, 1968, p. 734.
5 WUEST, W., Die Berechnung von Strömungsfeldern um dreidimensionale Körper in großer Höhe, DLR-FB, Göttingen, 1975.
6 DANG HONG TIEM, Ecoulements Moléculaires Libre et Presque Libre Autour d'un Obstacle Concave, Thèse de Doctorat de 3è Cycle, Université Paris 6, Paris, 1978.

Eingereicht am 23. 2. 1979, in revidierter Fassung am 21. 12. 1979

Anschrift: Dr. Dang Hong Tiem, Assistant, Département de Mathématiques, Faculté des Sciences de Reims, Moulin de la Housse, 51100 Reims, France (Ce travail a été effectué au Laboratoire de Mécanique théorique de l'Université Pierre et Marie Curie, Paris.)

ZAMM 60, 595–608 (1980)

W. H. Schmidt

Notwendige Optimalitätsbedingungen für Prozesse mit zeitvariablen Integralgleichungen in Banach-Räumen

Mit Hilfe eines abstrakten Modells von L. Bittner werden für Prozesse, die in Banach-Räumen durch Volterrasche oder Fredholmsche Integralgleichungen beschrieben werden, notwendige Optimalitätsbedingungen hergeleitet. Berücksichtigt wird eine Parameteroptimierung. Um Maximumprinzipien zu formulieren, nutzen wir die quellenmäßige Darstellung der Lösung einer Integralgleichung aus. Erörtert werden Existenz und Eigenschaften einer Resolvente für Integralgleichungen in Banach-Räumen.

By means of an abstract model of L. Bittner necessary optimality conditions are derived for processes governed by integral equations of Volterra or Fredholm types in Banach spaces. An optimization of parameters is considered. In order to formulate maximum principles we represent the solution of an integral equation by means of a resolvent operator. Existence and properties of resolvent operators for integral equations in Banach spaces are discussed.

С помощью общей модели Л. Биттнера приводятся необходимые условия оптимальности для процессов, описываемых интегральными уравнениями типов Вольтерра и Фредгольма в Банаховом пространстве. При этом рассматривается и оптимизация по параметру. Чтобы получить принципы максимума мы используем резольвентное представление решения интегрального уравнения. Исследуются также вопросы существования и свойств резольвента для интегральных уравнений в Банаховом пространстве.

1. Einleitung

Für gewisse optimale Prozesse in Banach-Räumen, bei denen die Prozeßgleichung eine Integralgleichung und das Funktional vom Lagrangeschen Typ ist, sollen in dieser Arbeit notwendige Optimalitätsbedingungen angegeben werden. Derartige Probleme mit Banach-Raum-wertigen Funktionen sind von Interesse, weil sie prinzipiell auch Prozesse mit verteilten Parametern zu behandeln gestatten. Durch Integralgleichungen beschriebene Prozesse in endlichdimensionalen Räumen findet man zuerst bei Butkovskij [4] und Vinokurov [13]. Wie Egorov [6] für Prozesse mit Differentialgleichungen in Banach-Räumen — ein Spezialfall der vorliegenden Problemstellung — zeigte, ist das Pontrjaginsche Maximumprinzip (siehe [3]) für solche Prozesse nicht gültig. Um ein Maximumprinzip herzuleiten, wird von ihm gefordert, daß der Steuerbereich innere Punkte besitzt. Wir betrachten Probleme mit freiem rechten Ende und werden ohne diese Voraussetzung auskommen.

Die Prozeßdauer und zusätzliche Parameter werden von uns als Steuergrößen aufgefaßt. Versucht man dagegen, Parameter durch Einführung neuer Gleichungen wie Zustandsvariable zu behandeln, so führt das auf Probleme, bei denen die Endpunkte der Trajektorien auf gewissen Mannigfaltigkeiten liegen, und man hat so den von Egorov betrachteten Fall erzeugt.

Bei der Herleitung notwendiger Optimalitätsbedingungen wird maßgeblich ein Modell von L. Bittner [1], [2] mit einer abstrakten Prozeßgleichung ausgenutzt. Aus den Optimalitätsbedingungen für dieses Modell gewinnt man „punktweise" Maximumprinzipien, indem man die Lösung von inhomogenen linearen Integralgleichungen in Banach-Räumen quellenmäßig darstellt. In vorbereitenden Betrachtungen wird die Resolvente zu einer linearen Integralgleichung definiert, und es werden Beziehungen zwischen Kern und Resolvente aufgedeckt.

Wir präzisieren die Aufgabenstellung: Es seien E, S und Q Banach-Räume und $J = [0, T]$ ein Zeitintervall. Gegeben seien Teilmengen U von S sowie P und P' von Q, dabei sei $P \subseteq P'$ und P' offen. Es bezeichne $\mathbb{R}$ bzw. $\mathbb{N}$ die Menge der reellen bzw. natürlichen Zahlen, $\overline{U}$ die Abschließung von U und $\tilde{C}(J, U)$ die Menge aller stückweise stetigen Funktionen $u: J \to U$, die bei 0 stetig und auf J linksseitig stetig sind. Entsprechend bezeichne $\tilde{C}(J, E^*)$ die Menge aller stückweise stetigen und linksseitig stetigen Funktionen von J in den Dualraum E^* zu E. $C(J, E)$ sei der Raum der über J definierten stetigen E-wertigen Funktionen. Es seien h und f stetige Abbildungen von $J \times P'$ bzw. $J \times J \times E \times \overline{U} \times P'$ in E und φ ein stetiges Funktional auf $J \times E \times \overline{U} \times P'$.

Folgende Probleme werden betrachtet:

$$\text{Minimiere} \int_0^T \varphi(t, x(t), u(t), p)\, dt \tag{1.1}$$

unter den Bedingungen $u \in \tilde{C}(J, U)$, $p \in P$, $T > 0$, $x \in C(J, E)$, $x(s) = h(s, p) + \int_0^s f(s, t, x(t), u(t), p)\, dt$, $s \in J$.

$$\text{Minimiere} \int_0^T \varphi(t, x(t), u(t), p)\, dt \tag{1.2}$$

unter den Bedingungen $u \in \tilde{C}(J, U)$, $p \in P$, $T > 0$, $x \in C(J, E)$, $x(s) = \int_0^T f(s, t, x(t), u(t), p)\, dt$, $s \in J$.

2. Einige Hilfssätze

Da die Steuerfunktionen stückweise stetig sein sollen, können alle auftretenden Integrale als Riemann-Integrale von Banach-Raum-wertigen Funktionen aufgefaßt werden. Grundlegende Rechenregeln dazu findet man bei Dieudonné [5]. Häufig wird der Satz von Fubini über die Vertauschbarkeit der Integrationsreihenfolge bei Doppelintegralen benutzt werden, dessen Voraussetzungen werden von den vorkommenden Integranden erfüllt. Für spätere Abschätzungen stellen wir nun einige Hilfssätze bereit.

Lemma 1: *Es seien Z ein* BANACH-*Raum und $a, b, c, \alpha, \beta, \delta, L$ reelle Zahlen mit $a < b < c$, $0 < \delta \ll 1$, $\alpha \leqq \beta, \alpha\beta \geqq 0$. Wenn $\gamma\colon [a, c] \times [0, \delta] \to \mathbb{R}$ eine im ersten Argument stetige Funktion ist, die der Bedingung $\int\limits_{b+\alpha\varepsilon}^{b+\beta\varepsilon} |\gamma(t, \varepsilon)|\, dt \leqq \varepsilon L$ für $0 < \varepsilon \leqq \delta$ genügt, so besteht für alle $g \in C([a, c], Z)$ die Beziehung*

$$\int\limits_{b+\alpha\varepsilon}^{b+\beta\varepsilon} g(t)\, \gamma(t, \varepsilon)\, dt = g(b) \int\limits_{b+\alpha\varepsilon}^{b+\beta\varepsilon} \gamma(t, \varepsilon)\, dt + o(\varepsilon);$$

$$\lim_{\varepsilon \to 0} \varepsilon^{-1} ||o(\varepsilon)|| = 0 .$$

Beweis: Der Beweis ergibt sich aus der Abschätzung

$$\left\|\int\limits_{b+\alpha\varepsilon}^{b+\beta\varepsilon} [g(t) - g(b)]\, \gamma(t, \varepsilon)\, dt\right\| \leqq \varepsilon L \max \{||g(t) - g(b)|| \mid b + \alpha\varepsilon \leqq t \leqq b + \beta\varepsilon\}$$

und der Stetigkeit von g.

Lemma 2: *Es sei Z ein* BANACH-*Raum, Ω eine kompakte Teilmenge des Raumes Q und k eine stetige Abbildung von $J \times E \times \overline{U} \times \Omega$ in Z. Für $x \in C(J, E)$ und $u \in \widetilde{C}(J, U)$ gilt dann*

$$\lim_{r \to 0} \sup \{||k(t, x(t) + e, u(t), p) - k(t, x(t), u(t), p)|| \mid p \in \Omega, t \in J, ||e|| \leqq r\} = 0 .$$

Beweis: u besitze n Unstetigkeitsstellen $0 < t_1 < \ldots < t_n < T$, es sei $t_0 = 0$ und $t_{n+1} = T$. Auf $[t_i, t_{i+1}]$ definieren wir stetige Funktionen $u_i\colon u_i(t) = u(t)$ für $t \neq t_i$ und $u_i(t_i) = \lim\limits_{t \to t_i, t > t_i} u(t)$. Durch einen indirekten, etwas schreibaufwendigen Beweis (siehe [11]) zeigt man für $i = 0, \ldots, n$

$$\lim_{r \to 0} \{||k(t, x(t) + e, u_i(t), p) - k(t, x(t), u_i(t), p)|| \mid p \in \Omega, t_i \leqq t \leqq t_{i+1}, ||e|| \leqq r\} = 0,$$

woraus die Aussage des Lemmas folgt.

Lemma 3: *Unter den Voraussetzungen von Lemma 2 gilt*

$$\lim_{q \to p,\, q \in \Omega} \sup \{||k(t, x(t), u(t), q) - k(t, x(t), u(t), p)|| \mid t \in J\} = 0 .$$

Beweis: Definiert man u_i wie beim Beweis von Lemma 2, so sind die Abbildungen $(t, p) \to k(t, x(t), u_i(t), p)$ auf den kompakten Mengen $[t_i, t_{i+1}] \times \Omega$ stetig und mithin gleichmäßig stetig, $i = 0, \ldots, n$. Das impliziert die Aussage des Lemmas.

Der Kegel

$$K(p, P) = \{q \in Q \mid \text{es existiert } \varepsilon(p, q), \text{ so daß } p + \varepsilon q \in P \text{ für } 0 \leqq \varepsilon \leqq \varepsilon(p, q)\}$$

heißt *Zelt der Menge P im Punkt p.*

Lemma 4: *Es seien Z ein* BANACH-*Raum und k eine stetige Abbildung von $J \times E \times \overline{U} \times P'$ in Z, es existiere die partielle* FRÉCHET-*Ableitung k_p (bezüglich $p \in Q$), und diese sei auf $J \times E \times \overline{U} \times P'$ stetig. Wenn $x \in C(J, E)$, $u \in \widetilde{C}(J, U)$, $p \in P$, $q \in K(p, P)$ und $\{\varepsilon_j\}$ eine Nullfolge positiver Zahlen ist, so gilt*

$$\lim_{j \to \infty} \sup \{\varepsilon_j^{-1} ||k(t, x(t), u(t), p + \varepsilon_j q) - k(t, x(t), u(t), p) - \varepsilon_j k_p(t, x(t), u(t), p)\, q|| \mid t \in J\} = 0 .$$

Beweis: Es sei $B(p)$ eine Kugel mit dem Mittelpunkt p, die in P' enthalten ist. Nach dem verallgemeinerten Mittelwertsatz gilt für alle $j \in \mathbb{N}$ mit $p + \varepsilon_j q \in B(p)$:

$$\varepsilon_j^{-1} ||k(t, x(t), u(t), p + \varepsilon_j q) - k(t, x(t), u(t), p) - k_p(t, x(t), u(t), p)\, \varepsilon_j q|| \leqq$$
$$\leqq ||q|| \sup_{t \in J} ||k_p(t, x(t), u(t), p + \vartheta_j(t)\, \varepsilon_j q) - k_p(t, x(t), u(t), p)||, \quad 0 < \vartheta_j(t) < 1.$$

Auf $[t_i, t_{i+1}]$ ersetzen wir wie beim Beweis von Lemma 2 u durch u_i. Die Abbildungen

$$(t, \varepsilon) \to k_p(t, x(t), u_i(t), p + \varepsilon q)$$

sind auf den kompakten Mengen $[t_i, t_{i+1}] \times [0, \varepsilon(p, q)]$ gleichmäßig stetig ($i = 0, \ldots, n$). Da J die endliche Vereinigung von Intervallen $[t_i, t_{i+1}]$ ist, ergibt sich Lemma 4.

3. Lineare Integralgleichungen in Banach-Räumen

Bei der Herleitung notwendiger Optimalitätsbedingungen für die Probleme (1.1) und (1.2) werden wir die quellenmäßige Darstellung der Lösung linearer Integralgleichungen in BANACH-Räumen ausnutzen.

Es sei $\mathcal{L}(E)$ der Raum der linearen beschränkten Operatoren auf E, und es bezeichne Π die Menge aller beschränkten Abbildungen B von $J \times J$ in $\mathcal{L}(E)$, für welche endlich viele Zahlen $0 < t_1 < \ldots < t_r < T$ existieren, so daß B auf den „Streifen" $J \times [0, t_1]$, $J \times]t_1, t_2]$, ..., $J \times]t_r, T]$ stetig ist.

Es sei nun $B \in \Pi$ und $K \geqq ||B(s, w)||$ für alle $s, w \in J$. Durch $(\hat{B}y)(s) = \int\limits_0^s B(s, w)\, y(w)\, dw$, $s \in J$, wird auf $C(J, E)$ ein linearer Operator $\hat{B}$ mit dem Spektralradius Null definiert. Daher besitzt die VOLTERRAsche Integralgleichung

$$z = \hat{B}z + g \tag{3.1}$$

im Raum $C(J, E)$ für jedes $g \in C(J, E)$ genau eine Lösung, diese kann mit dem gewöhnlichen Iterationsverfahren konstruiert werden.

Lemma 5: *Für jedes $n \in \mathbb{N}$ existiert eine Abbildung $R^n \in \Pi$, so daß die Iterationsfolge $z^0 = g$, $z^n = \hat{B}z^{n-1} + g$ die Darstellung*

$$z^n(s) = g(s) + \int_0^s R^n(s, w)\, g(w)\, dw \tag{3.2}$$

besitzt. Es ist $R^0(s, t) = 0$, und es bestehen die Rekursionsbeziehungen

$$R^{n+1}(s, t) = B(s, t) + \int_t^s B(s, w)\, R^n(w, t)\, dw \tag{3.3}$$

und

$$R^{n+1}(s, t) = B(s, t) + \int_t^s R^n(s, w)\, B(w, t)\, dw; \qquad s, t \in J. \tag{3.4}$$

Beweis: Durch Anwendung des Satzes von FUBINI erhält man (3.2) und (3.3). Um die Beziehung (3.4) zu beweisen, definiere man in Π:

$$V^0(s, t) = 0, \qquad V^{n+1}(s, t) = B(s, t) + \int_t^s V^n(s, w)\, B(w, t)\, dw, \qquad n \in \mathbb{N}.$$

Durch vollständige Induktion zeigt man die Beziehungen

$$R^{n+1}(s, t) = R^n(s, t) + \int_t^s \int_t^{t_n} \dots \int_t^{t_2} B(s, t_n)\, B(t_n, t_{n-1}) \dots B(t_2, t_1)\, dt_1 \dots dt_n$$

und

$$V^{n+1}(s, t) = V^n(s, t) + \int_t^s \int_{t_1}^s \dots \int_{t_{n-1}}^s B(s, t_n) \dots B(t_2, t_1)\, dt_n \dots dt_1$$

sowie

$$R^n(s, t) = V^n(s, t) \quad \text{für alle } s, t \in J \text{ und } n \in \mathbb{N}.$$

Lemma 6: *Es gilt die Abschätzung*

$$\|R^{n+1}(s, t) - R^n(s, t)\| \leqq \frac{K^{n+1}}{n!} |s - t|^n; \qquad s, t \in J, \qquad n \in \mathbb{N}.$$

Beweis: Für $n = 0$ ist die Ungleichung erfüllt; wird ihre Gültigkeit für $n = k$ vorausgesetzt, so folgt

$$\|R^{k+2}(s, t) - R^{k+1}(s, t)\| \leqq K \int_t^s \|R^{k+1}(w, t) - R^k(w, t)\|\, dw \leqq \frac{K^{k+2}}{(k+1)!} |s - t|^{k+1}.$$

Durch Anwendung von Lemma 6 ergibt sich:

Für alle $n, p \in \mathbb{N}$ gilt

$$\sup_{s,t \in J} \|R^{n+p}(s, t) - R^n(s, t)\| \leqq K \sum_{i=n}^{\infty} \frac{(KT)^i}{i!}.$$

In $\mathcal{L}(E)$ existiert mithin $R(s, t) = \lim_{n \to \infty} R^n(s, t)$ für jedes $s, t \in J$.

Satz 1: *Die so auf $J \times J$ definierte Abbildung R ist ein Element der Menge Π und erfüllt die Gleichungen*

$$R(s, t) = B(s, t) + \int_t^s B(s, w)\, R(w, t)\, dw, \tag{3.5}$$

$$R(s, t) = B(s, t) + \int_t^s R(s, w)\, B(w, t)\, dw; \qquad s, t \in J. \tag{3.6}$$

Beweis: Wegen $\|R(s, t) - R^n(s, t)\| = \lim_{m \to \infty} \|R^m(s, t) - R^n(s, t)\|$ gilt $\sup_{s,t \in J} \|R(s, t) - R^n(s, t)\| \leqq K \sum_{i=n}^{\infty} \frac{(KT)^i}{i!}$. Da die R^n auf $J \times J$ beschränkt sind, folgt hieraus die Beschränktheit von R. Wenn B auf den „Streifen" $J \times]t_i, t_{i+1}]$ stetig ist, so sind dort auch alle R^n stetig, und diese Eigenschaft vererbt sich auf R. Schließlich ist

$$\sup_{s,t \in J} \left\|R(s, t) - B(s, t) - \int_t^s B(s, w)\, R(w, t)\, dw\right\| \leqq (1 + KT) \sup_{s,t \in J} \|R(s, t) - R^n(s, t)\|.$$

Für $n \to \infty$ strebt die rechte Seite der Ungleichung nach 0. Dies zieht die Gleichung (3.5) nach sich. Gleichung (3.6) zeigt man analog.

Satz 2: *Die Lösung der Gleichung* (3.1) *läßt sich für jedes $g \in C(J, E)$ in der Form*

$$z(s) = g(s) + \int_0^s R(s, w)\, g(w)\, dw, \qquad s \in J, \tag{3.7}$$

darstellen. R wird Resolvente zur Gleichung (3.1) *genannt.*

Beweis: Aus der Abschätzung

$$||z - z^n|| \leqq \max_{s\in J} \int_0^s ||R(s, w) - R^n(s, w)||\, ||g(w)||\, dw \leqq ||g|| \sum_{i=n}^{\infty} \frac{(KT)^{i+1}}{i!}$$

folgt die Konvergenz der Iterationsfolge (3.2) gegen die durch (3.7) definierte Funktion z, mithin ist z die eindeutig bestimmte Lösung der Gleichung (3.1).

I bezeichne den identischen Operator, dann wird durch $U(s, t) = I + \int_t^s R(s, w)\, dw$, $s, t \in J$, eine Abbildung $U \in C(J \times J, \mathcal{L}(E))$ definiert. Die Gleichung (3.5) geht über in die Beziehung $U(s, t) = I + \int_0^s B(s, w)\, U(w, t)\, dw$, $s, t \in J$. Es sei $e \in E$ und $k \in C(J, E)$; in Gleichung (3.1) habe g die spezielle Gestalt $g(s) = e + \int_0^s k(t)\, dt$. Aus Satz 2 erhalten wir für die Lösung von (3.1) die Darstellung $z(s) = U(s, 0)\, e + \int_0^s U(s, t)\, k(t)\, dt$, $s \in J$. Falls B nicht vom ersten Argument abhängt, d. h. falls $B(s, t) = A(t)$ für alle $s, t \in J$ ist, folgt aus (3.6) die Beziehung $U(s, t) = I + \int_t^s U(s, w)\, A(w)\, dw$. In diesem Fall wird U *Evolutionsoperator zur Gleichung* (3.1) genannt. Die den Evolutionsoperator charakterisierenden Beziehungen implizieren die Gruppeneigenschaft $U(s, t) = U(s, w)\, U(w, t)$; $s, t, w \in J$.

Lemma 7: *Ist R die Resolvente zur Gleichung* (3.1) *und $\Lambda \in \tilde{C}(J, E^*)$, so ist die durch $\Psi_0(t) = \Lambda(t) + \int_t^T \Lambda(s)\, R(s, t)\, ds$ definierte Funktion die einzige stückweise stetige Lösung von*

$$\Psi(t) = \Lambda(t) + \int_t^T \Psi(s)\, B(s, t)\, ds\,. \tag{3.8}$$

Beweis: Mit Hilfe von (3.6) folgern wir, daß Ψ_0 eine Lösung von (3.8) ist:

$$\Psi_0(t) - \Lambda(t) - \int_t^T \Psi_0(s)\, B(s, t)\, ds = \int_t^T \Lambda(s)\, R(s,t)\, ds - \int_t^T \Lambda(s)\, B(s,t)\, ds - \int_t^T \int_s^T \Lambda(w)\, R(w,s)\, dw\, B(s,t) ds =$$

$$= \int_t^T \Lambda(s)\, [R(s, t) - B(s, t) - \int_t^s R(s, w)\, B(w, t)\, dw]\, ds = 0$$

für alle $t \in J$. Ist nun $\Psi \in \tilde{C}(J, E^*)$ eine beliebige Lösung von (3.8), so gilt

$$\int_t^T \Psi(s)\, R(s, t)\, ds = \int_t^T \int_t^s \Psi(s)\, B(s, w)\, R(w, t)\, dw\, ds + \int_t^T \Lambda(s)\, R(s, t)\, ds =$$

$$= \int_t^T \Psi(s)\, R(s, t)\, ds - \int_t^T \Psi(s)\, B(s, t)\, ds + \int_t^T \Lambda(s)\, R(s, t)\, ds$$

und folglich $\int_t^T \Psi(s)\, B(s, t)\, ds = \int_t^T \Lambda(s)\, R(s, t)\, ds$, $t \in J$, woraus $\Psi = \Psi_0$ folgt.

Die Untersuchung von Problem (1.2) wird uns zu FREDHOLMschen Integralgleichungen

$$z(s) = g(s) + \int_0^T B(s, t)\, z(t)\, dt\,, \qquad s \in J\,, \tag{3.9}$$

mit $B \in \Pi$ führen. Wir betrachten nur Gleichungen, die folgenden Bedingungen genügen:

(L) *Die Integralgleichung* (3.9) *sei für jedes $g \in C(J, E)$ eindeutig lösbar im Raum $C(J, E)$.*

(Q) *Für jedes $g \in C(J, E)$ lasse sich die Lösung von* (3.9) *mit Hilfe einer gewissen Abbildung $R \in \Pi$ quellenmäßig darstellen:* $z(s) = g(s) + \int_0^T R(s, t)\, g(t)\, dt$, $s \in J$.

Wir nennen R die *zur Gleichung* (3.9) *gehörende Resolvente.* Ist E endlichdimensional, so folgt (Q) bekanntlich aus (L). In BANACH-Räumen ist die Bedingung $\max_{s\in J} \int_0^T ||B(s, t)||\, dt < 1$ hinreichend dafür, daß (Q) und (L) erfüllt sind.

In diesem Fall läßt sich (3.9) mit dem gewöhnlichen Iterationsverfahren lösen, die Iterationsfolge kann man darstellen als $z^{(n)}(s) = g(s) + \int_0^T R^n(s, t)\, g(t)\, dt$, $s \in J$, mit $R^n \in \Pi$. Zu Lemma 6 und den Sätzen 1 und 2 analoge Überlegungen sichern die Existenz einer Resolvente und die quellenmäßige Darstellung der Lösung von (3.9).

Satz 3: *Wenn für die Gleichung* (3.9) *die Bedingungen* (L) *und* (Q) *erfüllt sind, so genügt die Resolvente R den Gleichungen*

$$R(s, t) = B(s, t) + \int_0^T B(s, w)\, R(w, t)\, dw \tag{3.10}$$

und

$$R(s, t) = B(s, t) + \int_0^T R(s, w)\, B(w, t)\, dw; \qquad s, t \in J\,. \tag{3.11}$$

Beweis: Auf $C(J, E)$ definieren wir lineare Integraloperatoren $\hat{B}$ und $\hat{R}$ mit den Kernfunktionen B bzw. R. Für die Lösung von (3.9) gilt somit $z = \hat{B}z + g$ und $z = \hat{R}g + g$ für alle $g \in C(J, E)$, hieraus folgen die Beziehungen $\hat{R} - \hat{B} = \hat{B}\hat{R}$ und $\hat{R} - \hat{B} = \hat{R}\hat{B}$. B sei auf den „Streifen" $J \times [0, t_1], J \times]t_1, t_2], \dots, J \times]t_r, T]$ stetig. Für jedes $e \in E$, $t \in J \setminus \{t_1, \dots, t_r\}$ und $0 < \varepsilon \ll 1$ definieren wir auf J eine stetige, stückweise lineare, E-wertige Funktion, die außerhalb des Intervalles $[t - \varepsilon, t + \varepsilon]$ verschwindet und im Punkt t den Wert e hat. Durch Anwendung von $\hat{B} - \hat{R} + \hat{B}\hat{R}$ auf diese Funktion folgt mit Hilfe von Lemma 1:

$$\varepsilon[B(s, t) - R(s, t) + \int_0^T B(s, w)\, R(w, t)\, dw]\, e + o(\varepsilon) = 0\,, \qquad s \in J\,.$$

Nach Division durch ε erhalten wir für $\varepsilon \to 0$ die Beziehung (3.10) für $t \neq t_1, \dots, t_r$. Aus der linksseitigen Stetigkeit von B und R folgt diese Beziehung für alle $s, t \in J$. Analog leitet man (3.11) her.

Korrolar: *$R - B$ ist auf $J \times J$ stetig, daher ist R auf den gleichen „Streifen" wie B stetig.*

Satz 4: *Es sei $B \in \Pi$. Wenn die Gleichungen* (3.10) *und* (3.11) *eine Lösung besitzen, so ist R Resolvente zur Gleichung* (3.9), *und die Gleichung* (3.9) *ist für jedes $g \in C(J, E)$ eindeutig lösbar, d. h. die Bedingungen* (L) *und* (Q) *sind erfüllt.*

Beweis: Mit der Lösung R von (3.10), (3.11) definieren wir $z(s) = g(s) + \int_0^T R(s, t)\, g(t)\, dt$, $s \in J$. Aus (3.10) folgt mit Hilfe des Satzes von FUBINI

$$\int_0^T R(s, t)\, g(t)\, dt + g(s) = \int_0^T B(s, w) \left[\int_0^T R(w, t)\, g(t)\, dt + g(w)\right] dw + g(s)\,; \qquad s \in J\,,$$

daher ist z eine Lösung von (3.9). Für eine beliebige Lösung $\tilde{z}$ von (3.9) gilt

$$\int_0^T R(s, t)\, \tilde{z}(t)\, dt = \int_0^T R(s, t)\, g(t)\, dt + \int_0^T\int_0^T R(s, w)\, B(w, t)\, dw\tilde{z}(t)\, dt =$$

$$= \int_0^T R(s, t)\, g(t)\, dt + \int_0^T R(s, t)\, \tilde{z}(t)\, dt - \int_0^T B(s, t)\, \tilde{z}(t)\, dt; \qquad s \in J\,,$$

und folglich $\tilde{z} = z$.

Lemma 8: *Wenn die Bedingungen* (L) *und* (Q) *erfüllt sind, $R \in \Pi$ die Resolvente zur Gleichung* (3.9) *bezeichnet und $\Lambda \in \tilde{C}(J, E^*)$ ist, so ist die durch*

$$\Psi_0(t) = \Lambda(t) + \int_0^T \Lambda(s)\, R(s, t)\, ds\,, \qquad t \in J\,, \tag{3.12}$$

definierte Funktion die einzige stückweise stetige Lösung von

$$\Psi(t) = \Lambda(t) + \int_0^T \Psi(s)\, B(s, t)\, ds\,, \qquad t \in J\,. \tag{3.13}$$

Verwendet man die adjungierten Operatoren B^* bzw. R^* zu B bzw. R, so können (3.12) *und* (3.13) *in der äquivalenten Form*

$$\Psi_0(t) = \Lambda(t) + \int_0^T R^*(s, t)\, \Lambda(s)\, ds \quad \text{bzw.} \quad \Psi(t) = \Lambda(t) + \int_0^T B^*(s, t)\, \Psi(s)\, ds\,; \qquad t \in J\,,$$

geschrieben werden.

Der Beweis ist wie bei Lemma 7 zu führen.

4. Prozesse mit Volterraschen Integralgleichungen und fixierter Prozeßdauer

Notwendige Optimalitätsbedingungen für das Problem (1.1) sollen aus den Ergebnissen von L. BITTNER für ein abstraktes Modell [1] hergeleitet werden. Dazu bezeichne $J = [0, T]$, $X = C(J, E)$, $W = \tilde{C}(J, U) \times P$, es sei $x \in X$ und $w = (u, p) \in W$. Durch

$$\hat{T}(x, w)(s) = h(s, p) + \int_0^s f(s, t, x(t), u(t), p)\, dt\,, \qquad s \in J\,, \quad \text{und} \quad F(x, w) = \int_0^T \varphi(t, x(t), u(t), p)\, dt$$

werden Abbildungen von $C(J, E) \times W$ in $C(J, E)$ bzw. $\mathbb{R}$ definiert. Die Aufgabe (1.1) ist dann äquivalent zu:

Minimiere $F(x, w)$ unter den Nebenbedingungen $x = \hat{T}(x, w)$, $w \in W$, $x \in X$.

Im weiteren werde vorausgesetzt, daß die partiellen FRÉCHET-Ableitungen h_p, φ_e, φ_p, f_e, f_p existieren und auf $J \times P'$ bzw. $J \times E \times \bar{U} \times P'$ bzw. $J \times J \times E \times \bar{U} \times P'$ stetig sind. Dies impliziert die in [1] an das abstrakte Modell gestellten Voraussetzungen: Aus der Stetigkeit der partiellen FRÉCHET-Ableitungen f_e und φ_e erhalten wir mit Hilfe von Lemma 2 (indem wir dort $Z = \mathscr{L}(E, C(J, E))$ und $k(t, e, u, p) = f_e(\cdot, t, e, u, p)$ bzw. $Z = E^*$, $k = \varphi_e$ setzen), daß die Abbildungen $\hat{T}$ und F stetige partielle FRÉCHET-Ableitungen bzgl. $x \in X$ besitzen.

Es ist

$$(\hat{T}_x(x, w)\Delta x)(s) = \int_0^s f_e(s, t, x(t), u(t), p)\, \Delta x(t)\, dt; \qquad s \in J,$$

und

$$F_x(x, w)\, \Delta x = \int_0^T \varphi_e(t, x(t), u(t), p)\, \Delta x(t)\, dt \qquad \text{für alle } \Delta x \in X.$$

Es sei nun $x_0 \in C(J, E)$, $u_0 \in \tilde{C}(J, U)$ und $p_0 \in P$ eine Lösung der Prozeßgleichung

$$x(s) = h(s, p) + \int_0^T f(s, t, x(t), u(t), p)\, dt; \qquad s \in J. \tag{4.1}$$

Wir setzen zur Abkürzung

$$f^*(s, t) = f(s, t, x_0(t), u_0(t), p_0), \qquad f_e^*(s, t) = f_e(s, t, x_0(t), u_0(t), p_0), \qquad f_p^*(t) = f_p(t, x_0(t), u_0(t), p_0),$$
$$\varphi^*(t) = \varphi(t, x_0(t), u_0(t), p_0), \qquad \varphi_e^*(t) = \varphi_e(t, x_0(t), u_0(t), p_0), \qquad \varphi_p^*(t) = \varphi_p(t, x_0(t), u_0(t), p_0).$$

Die zu (4.1) gehörende linearisierte Gleichung

$$\Delta x(s) = g(s) + \int_0^s f_e^*(s, t)\, \Delta x(t)\, dt, \qquad s \in J, \tag{4.2}$$

ist für jedes $g \in C(J, E)$ eindeutig im Raum $C(J, E)$ auflösbar.

Zu $w_0 = (u_0, p_0)$ werden *variierte Steuerungen* definiert. Dazu seien $v \in U$, $\tau \in \,]0, T]$ und $q \in K(p_0, P)$ beliebig (aber fest) und $\bar{\varepsilon}$ eine derartige Zahl, daß u_0 in $[\tau - \bar{\varepsilon}, \tau]$ stetig sowie $p_0 + \varepsilon q \in P$ für alle $0 \leqq \varepsilon \leqq \bar{\varepsilon}$ ist. Wir definieren $u_\varepsilon(t) = u_0(t)$ für $t \in J \setminus]\tau - \varepsilon, \tau]$, $u_\varepsilon(t) = v$ für $t \in \,]\tau - \varepsilon, \tau]$ und $p_\varepsilon = p_0 + \varepsilon q$. Es bezeichne W_0 die Menge $\{(u_\varepsilon, p_\varepsilon) \mid 0 \leqq \varepsilon \leqq \bar{\varepsilon}\}$.

Lemma 9: *W_0 ist eine Menge variierter Steuerungen für w_0 im Sinne von* BITTNER [1].

Beweis: Es ist zu zeigen, daß die Zahlen

$$\sup \{\|\hat{T}_x(x, w) - \hat{T}_x(x_0, w)\| + \|F_x(x, w) - F_x(x_0, w)\| \mid w \in W_0\}$$

nach 0 streben für $x \to x_0$. Wir setzen $x = x_0 + \Delta x$ und erhalten für alle $w \in W_0$ die Abschätzung

$$\|\hat{T}_x(x, w) - \hat{T}_x(x_0, w)\| + \|F_x(x, w) - F_x(x_0, w)\| \leqq$$
$$\leqq \int_0^T \|f_e(\cdot, t, x_0(t) + \Delta x(t), u_0(t), p_\varepsilon) - f_e(\cdot, t, x_0(t), u_0(t), p_\varepsilon)\|_{\mathcal{L}(E, C(J, E))}\, dt +$$
$$+ \int_0^T \|\varphi_e(t, x_0(t) + \Delta x(t), u_0(t), p_\varepsilon) - \varphi_e(t, x_0(t), u_0(t), p_\varepsilon)\|_{E^*}\, dt +$$
$$+ \int_{\tau-\varepsilon}^{\tau} \|f_e(\cdot, t, x_0(t) + \Delta x(t), v, p_\varepsilon) - f_e(\cdot, t, x_0(t), v, p_\varepsilon)\|_{\mathcal{L}(E, C(J, E))}\, dt +$$
$$+ \int_{\tau-\varepsilon}^{\tau} \|\varphi_e(t, x_0(t) + \Delta x(t), v, p_\varepsilon) - \varphi_e(t, x_0(t), v, p_\varepsilon)\|_{E^*}\, dt.$$

Die weitere Abschätzung folgt aus Lemma 2, wenn wir Z mit $\mathcal{L}(E, C(J, E))$ und $k(t, e, u, p)$ mit $f_e(\cdot, t, e, u, p)$ bzw. Z mit E^* und k mit φ_e und Ω mit $\{p_\varepsilon \mid 0 \leqq \varepsilon \leqq \bar{\varepsilon}\}$ identifizieren.

Es sei $\{\varepsilon_k\}$ eine Nullfolge mit $0 < \varepsilon_k \leqq \bar{\varepsilon}$, wir setzen $u_k := u_{\varepsilon_k}$ und $p_k := p_{\varepsilon_k}$ sowie $w_k = (u_k, p_k)$. Dann gilt

Lemma 10: *Die Folge $\{w_k\}$ ist konvergent gegen w_0 im Sinne von* BITTNER, *d. h.*

$$\lim_{k\to\infty} \{\|\hat{T}(x_0, w_k) - \hat{T}(x_0, w_0)\| + |F(x_0, w_k) - F(x_0, w_0)| + \|\hat{T}_x(x_0, w_k) - \hat{T}_x(x_0, w_0)\|$$
$$+ \|F_x(x_0, w_k) - F_x(x_0, w_0)\|\} = 0.$$

Beweis: Wir beschränken uns auf die Abschätzung des letzten Summanden, die übrigen Terme sind analog zu behandeln. Es ist

$$\|F_x(x_0, w_k) - F_x(x_0, w_0)\| \leqq \int_0^T \|\varphi_e(t, x_0(t), u_k(t), p_k) - \varphi_e^*(t)\|\, dt \leqq$$
$$\leqq T \sup \{\|\varphi_e(t, x_0(t), u_0(t), p_0 + \varepsilon_k q) - \varphi_e^*(t)\| \mid t \in J\} +$$
$$+ \varepsilon_k \sup \{\|\varphi_e(t, x_0(t), v, p_0 + \varepsilon q)\| + \|\varphi_e^*(t)\| \mid t \in J, 0 \leqq \varepsilon \leqq \bar{\varepsilon}\}.$$

Aus Lemma 3 mit $Z = E^*$, $k = \varphi_e$ und $\Omega = \{p_\varepsilon \mid 0 \leqq \varepsilon \leqq \bar{\varepsilon}\}$ sowie aus der Stetigkeit von φ_e folgt die gewünschte Beziehung.

Zur Herleitung notwendiger Optimalitätsbedingungen sind nach BITTNER [1] sog. *gemeinsame Richtungsgrenzwerte* zu berechnen. Im Fall VOLTERRAscher Integralgleichungen existieren diese jedoch nicht, weil die zu betrachtende Folge $\varepsilon_k^{-1}[\hat{T}(x_0, w_k) - \hat{T}(x_0, w_0)]$ zwar punktweise, jedoch nicht im Raum $C(J, E)$ konvergiert. Immerhin ist diese Folge aber beschränkt, denn es gilt

$$\varepsilon_k^{-1} \|\hat{T}(x_0, w_k) - \hat{T}(x_0, w_0)\| \leqq \varepsilon_k^{-1} \max_{s \in J} \|h(s, p_k) - h(s, p_0)\|_E +$$

$$+ \varepsilon_k^{-1} \int_0^T \|f(\cdot, t, x_0(t), u_0(t), p_k) - f^*(\cdot, t)\|_{C(J,E)} \, dt +$$

$$+ \varepsilon_k^{-1} \int_{\tau-\varepsilon_k}^{\tau} \|f(\cdot, t, x_0(t), v, p_k) - f^*(\cdot, t)\|_{C(J,E)} \, dt \, .$$

Nach Lemma 4 konvergiert der erste Summand der rechten Seite gegen $\|h_p(\cdot, p_0) q\|$ und der zweite gegen $\int_0^T \|\bar{f}_p^*(\cdot, t)\| \, dt$. Der dritte Summand ist beschränkt durch

$$\sup \{\|f^*(\cdot, t)\| + \|f(\cdot, x_0(t), v, p_0 + \varepsilon q)\| \mid \tau - \bar{\varepsilon} \leqq t \leqq \tau; 0 \leqq \varepsilon \leqq \varepsilon\} \, .$$

Es bezeichne $\delta F = \lim_{k\to\infty} \varepsilon_k^{-1}[F(x_0, w_k) - F(x_0, w_0)]$ und

$$\delta F T = \lim \varepsilon_k^{-1} F_x(x_0, w_0) [I - \hat{T}_x(x_0, w_0)]^{-1} [\hat{T}(x_0, w_k) - \hat{T}(x_0, w_0)] \, .$$

Lemma 11: *Es ist* $\delta F = \varphi(\tau, x_0(\tau), v, p_0) - \varphi^*(\tau) + \int_0^T \varphi_p^*(t) q \, dt$. *Bezeichnet R die Resolvente zur Gleichung* (4.2), *so gilt*

$$\delta F T = \int_0^T [\varphi_e^*(s) + \int_s^T \varphi_e^*(t) R(t, s) \, dt] [h_p(s, p_0) + \int_0^s f_p^*(s, w) \, dw] \, ds \, q +$$

$$+ \int_\tau^T [\varphi_e^*(s) + \int_s^T \varphi_e^*(t) R(t, s) \, dt] [f(s, \tau, x_0(\tau), v, p_0) - f^*(s, \tau)] \, ds \, . \tag{4.3}$$

Beweis: Auf die Darstellung von δF kommen wir in Lemma 15 zurück. Um $\delta F T$ zu berechnen, definieren wir $z_k = [I - \hat{T}_x(x_0, w_0)]^{-1} [\hat{T}(x_0, w_k) - \hat{T}(x_0, w_0)]$, also ist z_k die Lösung der Integralgleichung

$$z(s) = \int_0^s f^*(s, t) z(t) \, dt + h(s, p_k) - h(s, p_0) + \int_0^s [f(s, t, x_0(t), u_k(t), p_k) - f^*(s, t)] \, dt; \qquad s \in J \, .$$

Mit den Abkürzungen

$$\Delta h(s, p) = h(s, p) - h(s, p_0) + \int_0^s R(s, t) [h(t, p) - h(t, p_0)] \, dt \, ,$$

$$\Delta g(s, t, p) = f(s, t, x_0(t), u_0(t), p) - f^*(s, t) + \int_t^s R(s, w) [f(w, t, x_0(t), u_0(t), p) - f^*(w, t)] \, dw \, ,$$

$$\Delta f(s, t, y) = f(s, t, x_0(t), y, p_0) - f^*(s, t) + \int_t^s R(s, w) [f(w, t, x_0(t), y, p_0) - f^*(w, t)] \, dw \, ,$$

$$d(s, t, y, p) = f(s, t, x_0(t), y, p) - f(s, t, x_0(t), y, p_0) - f(s, t, x_0(t), u_0(t), p) + f^*(s, t) \, ,$$

$$r(s, t, y, p) = d(s, t, y, p) + \int_t^s R(s, w) \, d(w, t, y, p) \, dw$$

und der Resolventendarstellung von z_k ist

$$\delta F T = \lim_{k\to\infty} \varepsilon_k^{-1} \int_0^T \varphi_e^*(s) \left\{\Delta h(s, p_k) + \int_0^s \Delta g(s, t, p_k) \, dt + \int_0^s \Delta f(s, t, u_k(t)) \, dt + \int_0^s r(s, t, u_k(t), p_k) \, dt\right\} ds \, .$$

Aus Lemma 4 folgt

$$\lim_{k\to\infty} \varepsilon_k^{-1} \int_0^T \varphi_e^*(s) \, \Delta h(s, p_k) \, ds = \int_0^T \varphi_e^*(s) [h_p(s, p_0) + \int_0^s R(s, t) \, h_p(t, p_0) \, dt] \, ds \, q$$

und

$$\lim_{k\to\infty} \varepsilon_k^{-1} \int_0^T \varphi_e^*(s) \int_0^s \Delta g(s, t, p_k) \, dt \, ds = \int_0^T \varphi_e^*(s) \int_0^s [f_p^*(s, t) + \int_t^s R(s, w) \, f_p^*(w, t) \, dw] \, dt \, ds \, q \, .$$

Für $t \notin]\tau - \varepsilon_k, \tau]$ ist wegen $u_k(t) = u_0(t)$ auch $\Delta f(s, t, u_k(t)) = 0$. Aus der Abschätzung

$$\|\varepsilon_k^{-1} \int_0^\tau \varphi_e^*(s) \int_0^s \Delta f(s, t, u_k(t)) \, dt \, ds\| = \|\varepsilon_k^{-1} \int_{\tau-\varepsilon_k}^{\tau} \varphi_e^*(s) \int_{\tau-\varepsilon_k}^{\tau} \Delta f(s, t, v) \, dt \, ds\| \leqq$$

$$\leqq \varepsilon_k \sup \{\|\varphi_e^*(s)\| \mid \tau - \bar{\varepsilon} \leqq s \leqq \tau\} \sup \{\|\Delta f(s, t, v)\| \mid \tau - \leqq s, t \leqq \tau\}$$

ergibt sich $\lim_{k\to\infty} \varepsilon_k^{-1} \int_0^\tau \varphi_e^*(s) \int_0^s \Delta f(s, t, u_k(t))\, dt\, ds = 0$. Um $\lim_{k\to\infty} \varepsilon_k^{-1} \int_\tau^T \varphi_e^*(s) \int_0^s \Delta f(s, t, u_k(t))\, dt\, ds = \int_\tau^T \varphi_e^*(s)\ \Delta f(s, \tau, v)\, ds$ nachzuprüfen, wird die Abschätzung

$$||\varepsilon_k^{-1} \int_\tau^T \varphi_e^*(s) \left[\int_0^s \Delta f(s, t, u_k(t))\, dt - \varepsilon_k\ \Delta f(s, t, v)\right] ds|| \leqq \varepsilon_k^{-1} \int_\tau^T ||\varphi_e^*(s)|| \int_{\tau-\varepsilon_k}^\tau ||\Delta f(s, t, v) - \Delta f(s, \tau, v)||\, dt\, ds \leqq$$

$$\leqq T \sup \{||\varphi_e^*(s)|| \mid s \in J\} \sup \{||\Delta f(s, t, v) - \Delta f(s, \tau, v)|| \mid s \in J, \tau - \varepsilon_k \leqq t \leqq \tau\}$$

und die gleichmäßige Stetigkeit von $\Delta f(s, t, v)$ auf $J \times [\tau - \bar{\varepsilon}, \tau]$ benutzt.

Zusammen mit der vorhergehenden Überlegung liefert dies

$$\lim_{k\to\infty} \varepsilon_k^{-1} \int_0^T \varphi_e^*(s) \int_0^s \Delta f(s, t, u_k(t))\, dt\, ds =$$

$$= \int_\tau^T \varphi_e^*(s) \left[f(s, \tau, x_0(\tau), v, p_0) - f^*(s, \tau) + \int_\tau^s R(s, w) \left(f(w, \tau, x_0(\tau), v, p_0) - f^*(s, \tau)\right) dw\right] ds\,.$$

Schließlich kann $\left\|\varepsilon_k^{-1} \int_0^T \varphi_e^*(s) \int_0^s r(s, t, u_k(t), p_k)\, dt\, ds\right\|$ bis auf einen von k unabhängigen Faktor nach oben gegen

$$\sup \{||f(\cdot, t, x_0(t), v, p_k) - f(\ , t, x_0(t), v, p_0)|| + ||f(\cdot, t, x_0(t), u_0(t), p_k) - f^*(\cdot, t)|| \mid t \in J\}$$

abgeschätzt werden, gemäß Lemma 3 strebt die rechte Seite für $k \to \infty$ gegen Null.

Um die Resolvente R zu eliminieren, definieren wir auf E für jedes $t \in J$ ein Funktional

$$\Psi_0(t) = -\varphi_e^*(t) - \int_t^T \varphi_e^*(s)\ R(s, t)\, ds$$

und erhalten

Lemma 12: *Notwendig dafür, daß x_0, u_0, p_0 für den Prozeß* (1.1) *optimal sind, ist die Gültigkeit der Ungleichungen*

$$\varphi(t, x_0(t), v, p_0) - \int_t^T \Psi_0(s)\, f(s, t, x_0(t), v, p_0)\, ds \geqq \varphi^*(t) - \int_t^T \Psi_0(s)\, f^*(s, t)\, ds \tag{4.4}$$

und

$$\int_0^T \left\{\varphi_p^*(s) - \Psi_0(s) \left[h(s, p_0) + \int_0^s f_p^*(s, w)\, dw\right]\right\} ds\, q \geqq 0 \tag{4.5}$$

für jedes $q \in K(p_0, P)$, $v \in U$ und $t \in J$.

Beweis: Nach einem Satz von Bittner [1], dessen Voraussetzungen wegen der Lemmata 9 und 10 erfüllt sind, ist $\delta F + \delta F T \geqq 0$ notwendig für die Optimalität von x_0, $w_0 = (u_0, p_0)$ für alle Grenzwerte δF, $\delta F T$, die abgeleitet werden von einer Nullfolge $\{\varepsilon_k\}$ positiver Zahlen und einer gegen w_0 konvergenten Steuerfolge $\{w_k\}$, für welche $\varepsilon_k^{-1}\, ||\hat{T}(x_0, w_k) - \hat{T}(x_0, w_0)||$ beschränkt bleibt. Aus Lemma 11 folgen die Ungleichungen (4.4) und (4.5), wenn $q = 0$, $v \in U$ beliebig bzw. $v = u_0(t)$, $q \in K(p_0, P)$ beliebig gewählt wird und $t = \tau$ das Intervall $]0, T]$ durchläuft, wegen der Stetigkeit der Steuerfunktionen im Zeitpunkt 0 kann $t \in J$ geschrieben werden.

Gemäß Lemma 7 ist die eben definierte Funktion Ψ_0 in $\tilde{C}(J, E)^*$ die einzige Lösung von

$$\Psi(t) = \int_t^T \Psi(s)\, f_e^*(s, t)\, ds - \varphi_e^*(t)\,, \quad t \in J\,. \tag{4.6}$$

Wir nennen (4.6) die *zu x_0, u_0, p_0 gehörende adjungierte Gleichung.*

$H^1: J \times E \times \tilde{C}(J, E^*) \times U \times P \to \mathbb{R}$ werde definiert durch $H^1(t, e, \Psi, u, p) = \int_t^T \Psi(s)\, f(s, t, e, u, p)\, ds - \varphi(t, e, u, p)$.

Es bezeichne $H^{1*}(t)$, $H_e^{1*}(t)$, $H_p^{1*}(t)$ die Werte von H^1 bzw. der entsprechenden partiellen Fréchet-Ableitungen von H^1 zu den Argumenten $(t, x_0(t), \Psi_0(\cdot), u_0(t), p_0)$. Dann kann Lemma 12 umformuliert werden:

Satz 5: *Notwendig dafür, daß x_0, u_0, p_0 für den Prozeß* (1.1) *optimal sind, ist die Gültigkeit der Maximumbeziehung*

$$H^{1*}(t) = \max_{v \in U} H^1(t, x_0(t), \Psi_0(\cdot), v, p_0) \tag{4.7}$$

für alle $t \in J$ und das Bestehen der Ungleichung

$$\int_0^T [H_p^{1*}(t) + \Psi_0(t)\, h_p(t, p_0)]\, dt\, q \leqq 0 \tag{4.8}$$

für alle $q \in K(p_0, P)$.

5. Prozesse mit Differentialgleichungen

Prozesse mit Differentialgleichungen in BANACH-Räumen sind ein Spezialfall von Problemen des Typs

$$\int_0^T \varphi(t, x(t), u(t), p)\, dt = \min! \tag{5.1}$$

bezüglich $u \in \tilde{C}(J, U)$, $p \in P$, $x \in C(J, E)$,

$$x(s) = h(s, p) + \int_0^s \Phi(t, x(t), u(t), p)\, dt\,, \qquad s \in J\,.$$

Vorausgesetzt wird, daß Φ_e, Φ_p, φ_e, φ_p existieren und ebenso wie Φ und φ auf $J \times E \times \bar{U} \times P'$ stetig sind. Auf $J \times E \times E^* \times U \times P$ wird eine reellwertige Funktion H vermöge

$$H(t, e, \lambda, u, p) = \lambda \Phi(t, e, u, p) - \varphi(t, e, u, p)$$

definiert. Es sei x_0, u_0, p_0 eine Lösung der Prozeßgleichung. Wie oben bezeichnen $\Phi^*(t)$, $\Phi_e^*(t)$, $\varphi^*(t)$, $\varphi_e^*(t)$ die Werte von Φ, φ und den entsprechenden Ableitungen zu den Argumenten $(t, x_0(t), u_0(t), p_0)$. Wir nennen

$$\psi(s) = \int_s^T \psi(t)\, \Phi_e^*(t)\, dt - \int_s^T \varphi_e^*(t)\, dt\,, \qquad s \in J, \tag{5.2}$$

adjungierte Gleichung zu x_0, u_0, p_0. $H^*(t)$, $H_e^*(t)$, $H_p^*(t)$ sind die Werte von H, H_e und H_p zu den Argumenten $(t, x_0(t), \psi_0(t), u_0(t), p_0)$, dabei ist ψ_0 die Lösung von (5.2).

Satz 6: *Wenn* x_0, u_0, p_0 *für den Prozeß* (5.1) *optimal sind und* $\psi_0 \in C(J, E^*)$ *die Lösung der adjungierten Gleichung* (5.2) *ist, so besteht das Maximumprinzip*

$$H^*(t) = \max_{v \in U} H(t, x_0(t), \psi_0(t), v, p_0) \tag{5.3}$$

für alle $t \in J$ *und die Ungleichung*

$$\int_0^T [H_p^*(t) + H_e^*(t)\, h_p(t, p_0)]\, dt\, q \leqq 0 \tag{5.4}$$

für alle $q \in K(p_0, P)$.

Beweis: Da (5.1) einen Spezialfall von Prozeß (1.1) darstellt, sind die in Satz 5 formulierten Bedingungen (4.7) und (4.8) mit der Lösung Ψ_0 der entsprechenden Gleichung (4.6) erfüllt. Die durch $\psi_0(s) = \int_s^T \Psi_0(t)\, dt$, $s \in J$, definierte Funktion $\psi_0 \in C(J, E^*)$ ist dann die Lösung von (5.2). Wegen

$$H^1(t, e, \Psi, u, p) = \int_t^T \Psi(s)\, \Phi(t, e, u, p)\, ds - \varphi(t, e, u, p) = H(t, e, \psi(t), u, p)$$

folgt (5.3) aus der Bedingung (4.7). Für alle $q \in K(p_0, P)$ folgt aus (4.8) die Abschätzung

$$0 \geqq \int_0^T [H_p^{1*}(t) + \Psi_0(t)\, h_p(t, p_0)]\, dt\, q = \int_0^T H_p^*(t)\, dt + \int_0^T \left[\int_t^T \Psi_0(s)\, \Phi_e^*(t)\, ds - \varphi_e^*(t)\right] h_p(t, p_0)\, dt\, q =$$

$$= \int_0^T [H_p^*(t) + H_e^*(t)\, h_p(t, p_0)]\, dt\, q\,.$$

Im endlichdimensionalen Fall ist für autonome Systeme bekannt, daß die zu optimalen x_0, u_0, p_0 gebildete Funktion $H^*(t)$ auf J konstant ist (siehe [3]). Diese Aussagen werden für das Problem (5.1) verallgemeinert. Es wird vorausgesetzt, daß die partiellen Ableitungen h_t, φ_t und Φ_t existieren und stetig sind. M bezeichne die reelle Funktion

$$M(s) = H^*(s) + \int_s^T [H_t^*(w) + H_e^*(w)\, h_t(w, p_0)]\, dw; \qquad s \in J\,. \tag{5.5}$$

Lemma 13: *Auf* J *ist* M *stetig.*

Beweis: Zu zeigen ist die Stetigkeit von H^*. Unstetigkeitsstellen von H^* können nur die Unstetigkeitsstellen von u_0 sein. Angenommen, $\tau \in\,]0, T[$ ist eine Unstetigkeitsstelle von u_0 und $H^*(\tau - 0) > H^*(\tau + 0)$. Für alle $t' < \tau < t''$ mit $t'' - t' \ll 1$ ist dann $H(t'', x_0(t''), \psi_0(t''), u_0(t'), p_0) > H^*(t'')$, das widerspricht aber dem Maximumprinzip (5.3). Im Fall $H^*(\tau - 0) < H^*(\tau + 0)$ ist analog zu schließen.

Lemma 14: *M ist auf jedem Stetigkeitsintervall von* u_0 *konstant.*

Beweis: Angenommen, es existieren $\tau, \tau' \in J$ mit $\tau < \tau'$, so daß $M(\tau) \neq M(\tau')$ und u_0 auf $[\tau, \tau']$ stetig ist. Es bezeichne $K := \dfrac{|M(\tau') - M(\tau)|}{\tau' - \tau}$. Für jedes $m \in \mathbb{N}$ gibt es $\tau_m, \tau_m' \in [\tau, \tau']$ mit $\tau_m \neq \tau_m'$, $m|\tau_m - \tau_m'| < 1$ und $|M(\tau_m') - M(\tau_m)| > K|\tau_m' - \tau_m|$. Das bedeutet geometrisch (vgl. [3]), daß über $[\tau, \tau']$ jeder der Kurve M einbeschriebene Polygonzug wenigstens eine Kante besitzt, deren Anstieg dem Betrage nach nicht kleiner als K ist. Aus $\{\tau_m\}$, $\{\tau_m'\}$ können konvergente Teilfolgen ausgesondert werden, die wir wieder mit $\{\tau_m\}$ und $\{\tau_m'\}$ bezeichnen. Folglich existiert $\lim\limits_{m\to\infty} \tau_m = \lim\limits_{m\to\infty} \tau_m' = \tau_0 \in [\tau, \tau']$. Für $t_1, t_2 \in [\tau, \tau']$ gilt nach (5.3)

$$M(t_2) - M(t_1) \leqq H^*(t_2) - H(t_1, x_0(t_1), \psi_0(t_1), u_0(t_2), p_0) + \int_{t_2}^{t_1} [H_t^*(w) + H_e^*(w)\, h_t(w, p_0)]\, dw\,.$$

Die rechte Seite wird mit Hilfe des Mittelwertsatzes und des Mittelwertsatzes der Integralrechnung umgeformt:

$$M(t_2) - M(t_1) \leqq (t_2 - t_1) \left\{ \frac{dH}{dt} (t, x_0(t), \psi_0(t), u_0(t_2), p_0) \Big|_{t=\xi} - [H_t^*(\eta) + H_e^*(\eta)\, h_t(\eta, p_0)] \right\},$$

wobei ξ und η zwischen t_1 und t_2 liegen. Der Ausdruck in der Klammer verschwindet für $t_1 = t_2 = \xi = \eta = \tau_0$, er hängt stetig von t_1, t_2, ξ und η ab, weil u_0 in $[\tau, \tau']$ stetig ist. Für alle $t_1 \neq t_2$, für welche $|t_1 - \tau_0| + |t_2 - \tau_0|$ hinreichend klein ist, wird daher dieser Ausdruck dem Betrage nach kleiner als $K/2$. Für derartige t_1, t_2 ist also

$$M(t_2) - M(t_1) \leqq |t_2 - t_1| \frac{K}{2}$$

und, da t_1 und t_2 gleichberechtigt sind, auch

$$|M(t_2) - M(t_1)| \leqq |t_2 - t_1| \frac{K}{2}.$$

Für $m \geqq 1$ besitzen τ_m und τ'_m die von t_1, t_2 geforderten Eigenschaften, folglich ist $|M(\tau_m) - M(\tau'_m)| \leqq K/2\, |\tau_m - \tau'_m|$ im Widerspruch zur Konstruktion dieser Punkte.

Aus den Lemmata 13 und 14 folgt

Satz 7: *Die in* (5.5) *definierte Funktion M ist auf J konstant.*

6. Prozesse mit Fredholmschen Integralgleichungen und fixierter Prozeßdauer

Es sollen jetzt für das Problem (1.2) notwendige Optimalitätsbedingungen in Form von Maximumprinzipien hergeleitet werden. Es sei wieder $F(x, w) = \int_0^T \varphi(t, x(t), u(t), p)\, dt$, $w = (u, p)$, wir definieren $\hat{T}: C(J, E) \times W \to C(J, E)$ durch $\hat{T}(x, w)(s) = \int_0^T f(s, t, x(t), u(t), p)\, dt$, $s \in J$. Wir setzen wieder die Stetigkeit von f, f_e, f_p und φ, φ_e, φ_p voraus. Aus der Stetigkeit der partiellen FRÉCHET-Ableitung f_e und dem Lemma 2 (mit $Z = \mathcal{L}(E, C(J, E))$ und $k(t, e, u, p) = f_e(\cdot, t, e, u, p)$) erhält man, daß $\hat{T}$ partiell nach $x \in C(J, E)$ FRÉCHET-differenzierbar ist. Es gilt

$$\hat{T}_x(x, w)\, \Delta x = \int_0^T f_e(\cdot, t, x(t), u(t), p)\, \Delta x(t)\, dt\,.$$

Im weiteren sei x_0, u_0 und p_0 eine Lösung der Prozeßgleichung, d. h.

$$x_0(s) = \int_0^T f(s, t, x_0(t), u_0(t), p_0)\, dt\,, \qquad s \in J\,. \tag{6.1}$$

Es werden wieder die Abkürzungen

$$f^*(s, t) = f(s, t, x_0(t), u_0(t), p_0) \quad \text{und} \quad f_e^*(s, t) = f_e(s, t, x_0(t), u_0(t), p_0)\,, \qquad s, t \in J\,,$$

benutzt, analog wird $f_p^*(s, t)$ definiert.

Die zu (6.1) zugehörige Variationsgleichung lautet

$$z = g + \int_0^T f_e^*(\cdot, t)\, z(t)\, dt\,, \tag{6.2}$$

wobei g eine beliebige Funktion aus $C(J, E)$ ist. Die Gleichung (6.2) ist von der Art (3.9) und nicht immer lösbar. Wir setzen voraus, daß (6.2) der Bedingung (L) aus Kapitel 3 genügt, also für alle $g \in C(J, E)$ eindeutig lösbar ist.

Zu $w_0 = (u_0, p_0)$ sollen nun variierte Steuerungen definiert werden. Dazu geben wir uns Punkte $\tau_i \in\,]0, T]$, Vektoren $v_i \in U$ und $q \in K(p_0, P)$ sowie nichtnegative Zahlen α_i vor; $i = 1, \ldots, r$; dabei sei $\tau_1 \leqq \tau_2 \leqq \ldots \leqq \tau_r$. Wir setzen $l_i = \alpha_i + \ldots + \alpha_j$, falls $\tau_i = \ldots = \tau_j < \tau_{j+1}$ und $l_i = \alpha_i + \ldots + \alpha_r$, falls $\tau_i = \ldots = \tau_r$ ist, sowie $J_i(\varepsilon) =]\tau_i - \varepsilon l_i, \tau_i - \varepsilon(l_i - \alpha_i)]$, $i = 1, \ldots, r$. Hierbei sei $0 \leqq \varepsilon \leqq \bar{\varepsilon}$ und $\bar{\varepsilon} > 0$ so klein, daß u_0 in allen $J_i(\varepsilon)$ stetig ist, die Intervalle $J_i(\varepsilon)$ paarweise disjunkt und alle in J enthalten sind, sowie $p_0 + \varepsilon q \in P$ für $0 \leqq \varepsilon \leqq \bar{\varepsilon}$ gilt. Wir definieren $u_\varepsilon(t) = v_i$ für $t \in J_i(\varepsilon)$, $u_\varepsilon(t) = u_0(t)$ für $t \in J \setminus \bigcup_{i=1}^r J_i(\varepsilon)$ und $p_\varepsilon = p_0 + \varepsilon q$. Dann sei

$$W_0 = \{(u_\varepsilon, p_\varepsilon) \mid 0 \leqq \varepsilon \leqq \bar{\varepsilon}\}\,.$$

Lemma 14: *W_0 ist eine Menge variierter Steuerungen.*

Beweis: Wie bei Lemma 9 ist zu zeigen, daß $\hat{T}_x$ und F_x bezüglich x stetig bei x_0 sind, und zwar gleichmäßig in $w \in W_0$. Es sei $\Omega = \{p_\varepsilon \mid 0 \leqq \varepsilon \leqq \bar{\varepsilon}\}$. Für $w \in W_0$ ist

$$\|\hat{T}_x(x, w) - \hat{T}_x(x_0, w)\| \leqq \int_0^T \|f_e(\cdot, t, x(t), u_\varepsilon(t), p_\varepsilon) - f_e(\cdot, t, x_0(t), u_\varepsilon(t), p_\varepsilon)\|\, dt \leqq$$
$$\leqq T \sup \{\|f_e(\cdot, t, x(t), u_0(t), p) - f_e(\cdot, t, x_0(t), u_0(t), p)\| \mid t \in J, p \in \Omega\} +$$
$$+ \bar{\varepsilon} \sum_{i=1}^r \alpha_i \sup \{\|f_e(\cdot, t, x(t), v_i, p) - f_e(\cdot, t, x_0(t), v_i, p)\| \mid t \in J, p \in \Omega\}\,.$$

Die rechte Seite hängt von $w \in W_0$ nicht mehr ab und strebt für $x \to x_0$ gegen 0 (Lemma 2). Analog behandelt man F_x.

$\{\varepsilon_k\}$ sei eine Nullfolge mit $0 \leq \varepsilon_k \leq \bar{\varepsilon}$, $k \in \mathbb{N}$. Wir setzen $u_k = u_{\varepsilon_k}$, $p_k = p_{\varepsilon_k}$ und $w_k = (u_k, p_k)$. In gleicher Weise wie bei Lemma 10 erhält man, daß die Steuerfolge $\{w_k\}$ im Sinne von Bittner gegen w_0 konvergent ist.

Lemma 15: *Die gemeinsamen Richtungsgrenzwerte*

$$\delta F = \lim_{k\to\infty} \varepsilon_k^{-1}[F(x_0, w_k) - F(x_0, w_0)], \qquad \delta T = \lim_{k\to\infty} \varepsilon_k^{-1}[\hat{T}(x_0, w_k) - \hat{T}(x_0, w_0)]$$

existieren, und es ist

$$\delta F = \sum_{i=1}^{r} \alpha_i[\varphi(\tau_i, x_0(\tau_i), v_i, p_0) - \varphi^*(\tau_i)] + \int_0^T \varphi_p^*(t)\, dt\, q\,,$$

$$\delta T = \sum_{i=1}^{r} \alpha_i[f(\cdot, \tau_i, x_0(\tau_i), v_i, p_0) - f^*(\cdot, \tau_i) + \int_0^T f_p^*(\cdot, t)\, dt\, q\,.$$

Beweis: Wir zeigen die Darstellung von δF, analog erhält man δT:

$$\delta F = \lim_{k\to\infty} \varepsilon_k^{-1} \int_0^T [\varphi(t, x_0(t), u_k(t), p_0) - \varphi^*(t)]\, dt +$$

$$+ \lim_{k\to\infty} \varepsilon_k^{-1} \int_0^T [\varphi(t, x_0(t), u_0(t), p_k) - \varphi^*(t)]\, dt +$$

$$+ \lim_{k\to\infty} \varepsilon_k^{-1} \int_0^T [\varphi(t, x_0(t), u_k(t), p_k) - \varphi(t, x_0(t), u_k(t), p_0) - \varphi(t, x_0(t), u_0(t), p_k) + \varphi^*(t)]\, dt\,.$$

Die variierte Steuerung u_k ist nur auf den Intervallen $J_i(\varepsilon_k)$ von u_0 verschieden. Nach Lemma 1 ist daher der erste Summand der rechten Seite $\sum_{i=1}^{r} \alpha_i[\varphi(\tau_i, x_0(\tau_i), v_i, p_0) - \varphi^*(\tau_i)]$. Aus Lemma 4 erhält man für den zweiten Summanden $\int_0^T \varphi_p^*(t)\, dt\, q$. Schließlich ist

$$\varepsilon_k^{-1} \left\|\int_0^T [\varphi(t, x_0(t), u_k(t), p_k) - \varphi(t, x_0(t), u_k(t), p_0) - \varphi(t, x_0(t), u_0(t), p_k) + \varphi^*(t)]\, dt\right\| \leq$$

$$\leq \sum_{i=1}^{r} \alpha_i\{\sup_{t\in J} \|\varphi(t, x_0(t), v_i, p_k) - \varphi(t, x_0(t), v_i, p_0)\| + \sup_{t\in J} \|\varphi(t, x_0(t), u_0(t), p_k) - \varphi(t, x_0(t), u_0(t), p_0)\|\}\,,$$

dieser Ausdruck strebt für $k \to \infty$ gemäß Lemma 3 gegen 0.

Satz 7: *Wenn das Zelt $K(p_0, P)$ konvex ist, bildet die Menge $\{(\delta F, \delta T)\}$ aller gemeinsamen Richtungsgrenzwerte einen konvexen Kegel.*

Beweis: Da die Menge der gemeinsamen Richtungsgrenzwerte immer einen Kegel bildet, ist nur die Konvexität zu zeigen. Diese folgt aus der Konstruktion der variierten Steuerungen (mit beliebigen Störstellen τ_i und beliebigen Vektoren $v_i \in U$) und der Konvexität von $K(p_0, P)$.

Satz 7 ist von Bedeutung, wenn bei Problem (1.2) zusätzlich gewisse Phasenbeschränkungen erfaßt werden sollen (siehe [2]).

Wir geben jetzt ein erstes Optimalitätskriterium an:

Satz 8: *Für jedes $\tau \in J$ und $v \in U$ sei $\chi_0 = \chi_0(\cdot, \tau, v)$ die Lösung von*

$$\chi(s) = \int_0^T f_e^*(s, t)\, \chi(t)\, dt - f(s, \tau, x_0(\tau), v, p_0)\,, \qquad s \in J\,, \tag{6.3}$$

im Raum $C(J, E)$ und

$$\mathcal{H}(\tau, v) = \int_0^T \varphi_e^*(t)\, \chi_0(t, \tau, v)\, dt - \varphi(\tau, x_0(\tau), v, p_0)\,.$$

Für die Optimalität von x_0, u_0, p_0 ist es notwendig, daß die Maximumbeziehung $\mathcal{H}(\tau, u_0(\tau)) = \max_{v\in U} \mathcal{H}(\tau, v)$ für alle $\tau \in J$ erfüllt ist.

Beweis: Die Voraussetzung (L) sichert die eindeutige Lösbarkeit von (6.3). Notwendig für die Optimalität von x_0 und $w_0 = (u_0, p_0)$ ist $F_x(x_0, w_0) [I - T_x(x_0, w_0)]^{-1}\, \delta T + \delta F \geq 0$ für alle gemeinsamen Richtungsgrenzwerte (siehe [1]). Es werden nun nur solche δT, δF betrachtet, die mittels variierter Steuerungen mit einer spikeförmigen Abänderung von u_0 im Intervall $]\tau - \varepsilon, \tau]$ durch einen Vektor $v \in U$ erzeugt werden, der Parameter wird nicht variiert, d. h. $q = 0$, $\alpha = 1$, $\tau_1 = \tau$, $v_1 = v$. Satz 8 ergibt sich dann aus Lemma 15 und der Beziehung (6.3).

Das in Satz 8 formulierte Kriterium hat einen entscheidenden Nachteil: Für jedes $\tau \in J$ und $v \in U$ ist die Gleichung (6.3) mit jeweils anderer rechter Seite zu lösen. Wir wollen deshalb $[I - \hat{T}_x(x_0, w_0)]^{-1}$ mittels einer Resolvente eliminieren und setzen dazu voraus, daß die Variationsgleichung (6.2) der Bedingung (Q) aus Kapitel 3 genügt, ihre Lösung sich also für alle $g \in C(J, E)$ mit einer Resolvente $R \in \Pi$ quellenmäßig darstellen läßt. Hinreichende Bedingungen für das Zutreffen von (Q) wurden oben angegeben. Die Gleichung

$$\Psi(t) = \int_0^T \Psi(s)\, f_e^*(s, t)\, ds - \varphi_e^*(t)\,, \qquad t \in J\,, \tag{6.4}$$

heißt *adjungierte Gleichung zu* x_0, u_0, p_0. Auf $J \times E \times \widetilde{C}(J, E^*) \times U \times P$ definieren wir H^2 durch

$$H^2(t, e, \Psi, v, p) = \int_0^T \Psi(s)\, f(s, t, e, u, p)\, ds - \varphi(t, e, u, p)\,.$$

Satz 9: *Es seien* x_0, u_0, p_0 *optimal für den Prozeß* (1.2) *und* $\Psi_0 \in \widetilde{C}(J, E^*)$ *die Lösung der adjungierten Gleichung* (6.4). $H^{2*}(t)$ *bzw.* $H_p^{2*}(t)$ *seien die Werte von* H^2 *bzw.* H_p^2 *zu den Argumenten* $(t, x_0(t), \Psi_0, u_0(t), p_0)$. *Dann bestehen die Maximumbeziehung*

$$H^{2*}(t) = \max_{v \in U} H^2(t, x_0(t), \Psi_0, v, p_0) \quad \text{für jedes } t \in J \tag{6.5}$$

und die Ungleichung

$$\int_0^T H_p^{2*}(t)\, dt\, q \leqq 0 \quad \text{für alle } q \in K(p_0, P)\,. \tag{6.6}$$

Beweis: Wegen der Optimalität von x_0, $w_0 = (u_0, p_0)$ ist $F_x(x_0, w_0)\,[I - \hat{T}_x(x_0, w_0)]^{-1}\,\delta T + \delta F \geqq 0$. Die Lösung der Integralgleichung $z = \hat{T}_x(x_0, w_0)\, z + \delta T$ stellen wir quellenmäßig dar, außerdem variieren wir Steuerfunktion ($q = 0$) und Parameter ($v = u_0(\tau)$) getrennt. Damit erhalten wir

$$\int_0^T [\varphi_e^*(t) + \int_0^T \varphi_e^*(s)\, R(s, t)\, ds]\, [f(t, \tau, x_0(\tau), v, p_0) - f^*(t, \tau)]\, dt + \varphi(\tau, x_0(\tau), v, p_0) - \varphi^*(\tau) \geqq 0$$

und

$$\int_0^T [\varphi_e^*(t) + \int_0^T \varphi_e^*(s)\, R(s, t)\, ds] \int_0^T f_p^*(t, w)\, dw\, dt\, q + \int_0^T \varphi_p^*\, dt\, q \geqq 0\,, \qquad \tau \in J\,, \quad v \in U\,, \quad q \in K(p_0, P)\,.$$

Nach Lemma 8 ist $\Psi_0(t) = -\varphi_e^*(t) - \int_0^T \varphi_e^*(s)\, R(s, t)\, ds$, $t \in J$, die Lösung der adjungierten Gleichung (6.4), mit der Funktion H^2 folgen (6.5) und (6.6).

7. Prozesse mit variabler Endzeit

Wir sind jetzt in der Lage, die Probleme (1.1), (1.2) und (5.1) mit variabler Endzeit T zu studieren, indem wir solche Aufgaben auf Probleme mit fixierter Prozeßdauer transformieren und die Endzeit als reellen Parameter auffassen. Vorausgesetzt wird die Existenz stetiger FRÉCHET-Ableitungen φ_t, Φ_t, f_s, f_t.

Es seien x_0, u_0, p_0 und die Prozeßdauer T_0 eine Lösung des Problems (1.1). In der Definition der adjungierten Gleichung (4.6) und der Funktion H^1 ist T durch T_0 zu ersetzen.

Satz 10: *Wenn* x_0, u_0, p_0 *und* T_0 *eine optimale Lösung von* (1.1) *bilden, so sind die Maximumbeziehung* (4.7), *die Ungleichung* (4.8) *mit* $T = T_0$ *und die Gleichung*

$$\int_0^{T_0} \{H^{1*}(t) + H_t^{1*}(t) \cdot t + \Psi_0(t)\, [h_t(t, p_0) + \int_0^t f_s^*(t, w)\, dw]\, t\}\, dt = 0 \tag{7.1}$$

erfüllt, dabei ist Ψ_0 *die Lösung der zu* x_0, u_0, p_0 *und* T_0 *gehörenden Gleichung* (4.6).

Beweis: Vermöge $t \to t/T$ wird das Intervall $[0, T]$ auf $[0, 1]$ transformiert, wir setzen $\hat{x}(t) = x(Tt)$, $\hat{u}(t) = u(Tt)$. Die Aufgabe (1.1) geht dabei über in

$$\int_0^1 \varphi(Tt, \hat{x}(t), \hat{u}(t), p)\, T\, dt = \min! \qquad \text{bezüglich } \hat{x} \in C([0, 1], E),\ \hat{u} \in \widetilde{C}([0, 1], U),\ p \in Q,\ T > 0\,,$$

$$x(s) = h(Ts, p) + \int_0^s f(Ts, Tt, x(t), u(t), p)\, T\, dt\,, \quad s \in [0, 1].$$

Für dieses Problem sind $\hat{x}_0$, $\hat{u}_0$ und die Parameter p_0, T_0 dann und nur dann optimal, wenn x_0, u_0, p_0 und die Zeit T_0 für den Prozeß (1.1) optimal sind. $\hat{\Psi}_0 \in \widetilde{C}([0, 1], E^*)$ sei die Lösung von

$$\Psi(t) = \int_t^1 \Psi(s)\, f_e(T_0 s, T_0 t, \hat{x}_0(t), \hat{u}_0(t), p_0)\, T_0\, ds - \varphi_e(T_0 t, \hat{x}_0(t), \hat{u}_0(t), p_0)\, T_0\,, \qquad 0 \leqq t \leqq 1\,.$$

Um Optimalitätsbedingungen für den transformierten Prozeß zu gewinnen, notieren wir die zugehörige Ungleichung (4.5) mit $(0, 1)$ und $(0, -1) \in K(p_0, P) \times \mathbb{R}$:

$$\int_0^1 \hat{\Psi}_0(s) \Big\{h_s(T_0 s, p_0)\, s + \int_0^s [f(T_0 s, T_0 t, \hat{x}_0(t), \hat{u}_0(t), p_0) + f_s(T_0 s, T_0 t, \hat{x}_0(t), \hat{u}_0(t), p)\, T_0 s +$$

$$+ f_t(T_0 s, T_0 t, \hat{x}_0(t), \hat{u}_0(t), p_0)\, T_0 t]\, dt\Big\}\, ds =$$

$$= \int_0^1 [\varphi(T_0 t, \hat{x}_0(t), \hat{u}_0(t), p_0) + \varphi_t(T_0 t, \hat{x}_0(t), \hat{u}_0(t), p_0)\, T_0 t]\, dt\,.$$

Die durch $\Psi_0(t) = \frac{1}{T_0} \hat{\Psi}_0\left(\frac{t}{T_0}\right)$ auf $[0, T_0]$ definierte Funktion Ψ_0 ist die Lösung der adjungierten Gleichung zu Problem (1.1). Der Übergang von [0, 1] zu $[0, T_0]$ durch eine lineare Transformation liefert die Gleichung

$$\int_0^{T_0} \Psi_0(s) \left\{h_s(s, p_0)\, s + \int_0^s [f^*(s, t) + f_s^*(s, t)\, s + f_t^*(s, t)\, t]\, dt\right\} ds = \int_0^{T_0} [\varphi^*(t) + \varphi_t^*(t)\, t]\, dt ,$$

woraus 7.1 folgt.

Satz 11: *Es sei x_0, u_0, p_0 und die Zeit T_0 eine optimale Lösung des Problems* (5.1) *und ψ_0 die Lösung der adjungierten Gleichung*

$$\psi(s) = \int_s^{T_0} \psi(t)\, \Phi_e^*(t)\, dt - \int_s^{T_0} \varphi_e^*(t)\, dt , \qquad s \in J .$$

Dann sind das Maximumprinzip, die Ungleichung (5.3) *mit $T = T_0$ und die Gleichung*

$$H^*(t) = \int_{T_0}^{t} [H_t^*(s) + H_e^*(s)\, h_t(s, p_0)]\, ds , \qquad 0 \leqq t \leqq T_0 , \tag{7.2}$$

erfüllt.

Beweis: Aus Satz 10 erhält man zur Charakterisierung der optimalen Prozeßdauer T_0 die Gleichung

$$\int_0^{T_0} [H^*(t) + H_t^*(t)\, t + H_e^*(t)\, h_t(t, p_0)\, t]\, dt = 0 .$$

Durch partielle Integration ergibt sich

$$\int_0^{T_0} \left\{H^*(t) + \int_t^{T_0} [H_t^*(s) + H_e^*(s)\, h_t(s, p_0)]\, ds\right\} dt = 0 ,$$

was zusammen mit Satz 7 gerade (7.2) liefert.

Satz 12: *Es sei x_0, u_0, p_0 und die Zeit T_0 eine optimale Lösung von* (5.2) *und Ψ_0 die Lösung der adjungierten Gleichung*

$$\Psi(t) = \int_0^{T_0} \Psi(s)\, f_e^*(s, t)\, ds - \varphi_e^*(t) , \qquad 0 \leqq t \leqq T_0 .$$

Dann sind die Maximumbeziehung (6.5) *und die Ungleichung* (6.6) *mit $T = T_0$ sowie die Gleichung*

$$\int_0^{T_0} \left[H^{2*}(t) + H_t^{2*}(t)\, t + \int_0^{T_0} \Psi_0(s)\, f_s^*(s, t)\, s\, ds\right] dt = 0 \tag{7.3}$$

erfüllt.

Satz 12 wird wie Satz 10 bewiesen, indem nach einer Zurückführung auf einen Prozeß mit fixierter Prozeßdauer aus (6.6) für den Parameter T_0 eine Bedingung abgeleitet wird, diese ist dann zu (7.3) umzuformen.

8. Die Methode von Krylow und Tschernousko

Zur Berechnung der optimalen Steuerung aus den Maximumprinzipien wird ein in [9] dargestelltes heuristisches Verfahren von KRYLOV und TSCHERNOUSKO verallgemeinert. Bei der Darstellung beschränken wir uns auf den Prozeß (1.1) und sehen von Parametern ab:

(i) *Wähle eine Dispatchersteuerung $u_1 \in \tilde{C}(J, U)$.*

(ii) *Ermittle die Lösung x_1 der Prozeßgleichung $x(s) = h(s) + \int_0^s f(s, t, x(t), u_1(t))\, dt$, $s \in J$.*

(iii) *Bestimme die Lösung Ψ_1 der adjungierten Gleichung*

$$\Psi(t) = \int_t^T \Psi(s)\, f_e(s, t, x_1(t), u_1(t))\, ds - \varphi_e(t, x_1(t), u_1(t)) , \qquad t \in J .$$

(iv) *Bestimme $u_2(t)$ als Lösung des Optimierungsproblems*

$$H^1(t, x_1(t), \Psi_1, v) = \max_{v \in U}! , \qquad t \in J .$$

Falls u_2 und das zugehörige x_2 zu einer Abnahme des Zielfunktionals führen, wiederhole die Prozedur mit u_2 anstelle von u_1.

Konvergenzaussagen können nicht getroffen werden. Im Fall linearer Prozesse ist die Prozedur nach einem Schritt stationär, d. h. $u_i = u_{i+1}$ für $i \geqq 2$.

Um die Optimierungsaufgaben (iv) durch endlichdimensionale Probleme näherungsweise zu lösen, wird vorgeschlagen, „dem Problem möglichst angepaßte" Ansatzfunktionen $\mu_1, \ldots, \mu_m \in \tilde{C}(J, U)$ zu wählen und nur Steuerungen u mit $u(t) = \sum_{i=1}^m a_i(t)\, \mu_i(t)$, $a_i(t) \in \mathbb{R}$; $i = 1, \ldots, m$; zuzulassen. Falls der Steuerbereich U konvex ist, wird durch die Forderung $a_i(t) \geqq 0$, $\sum_{i=1}^m a_i(t) = 1$ und $a_i \in \tilde{C}(J, \mathbb{R})$; $i = 1, \ldots, m$; $t \in J$; gesichert, daß die $u = \sum_{i=1}^m a_i \mu_i$ zulässig sind.

Literatur

1 BITTNER, L., On optimal control of processes governed by abstract functional, integral and hyperbolic differential equations, Math. Operationsforsch. Statist. **6** (1975), 107–134.
2 BITTNER, L., Necessary optimality conditions for a model of optimal control processes, Banach Center Publ. Warschau 1976 **1**, 25–32.
3 BOLTJANSKIJ, W. G., Mathematische Methoden der optimalen Steuerung (Übersetzung aus dem Russischen), Akad. Verlagsgesellschaft, Leipzig 1971.
4 BUTKOVSKIJ, A. G. (Бутковский, А. Г.) Теория оптимального управления системы с распределенными параметрами, Наука, Москва 1965.
5 DIEUDONNÉ, J., Grundzüge der modernen Analysis. Band 1 (Übersetzung aus dem Englischen), Deutscher Verlag der Wissenschaften, Berlin 1971.
6 EGOROV, J. V. (Егоров, Ю. В.), Необходимые условия оптимальности управления в Банаховом пространстве, Мат. Сборник **64**, I (1964), 79–101.
7 KANTOROWITSCH, L. W.; AKILOW, G. P., Funktionalanalysis in normierten Räumen (Übersetzung aus dem Russischen), Akademie-Verlag, Berlin 1964.
8 KREIN, S. G. (Крайн, С. Г.), Линейные дифференциальные уравнения в Банаховом пространстве, Наука, Москва 1967.
9 MOISEJEW, N. N. (Моисеев, Н. Н.), Численные методы в теории оптимальных систем, Наука, Москва 1971.
10 RIESZ, F.; SZ.-NAGY, B., Vorlesungen über Funktionalanalysis, Deutscher Verlag der Wissenschaften, Berlin 1973.
11 SCHMIDT, W. H., Optimalitätsbedingungen für Prozesse mit zeitvariablen Differential- und Integralgleichungen in Banachräumen, Dissertation A, Greifswald 1976.
12 SPREMAN, K., Optimierung verschiedener Steuerungsprobleme mit einem funktionalanalytischen Maximumprinzip, ZAMM **54**, 293–302 (1974).
13 VINOKUROV, V. R. (Винокуров, В. Р.), Оптимальное управление процессами описываемыми интегральными уравнениями. I, II, III, Известия высш. учебн. завед., сер. мат. **1967**, № 7, 8, 9.

Eingereicht am 17. 3. 1980

Anschrift: Dr. W. H. SCHMIDT, Universität Greifswald, Sektion Mathematik, L.-Jahnstraße 15a, DDR-2200 Greifswald

ZAMM 60, 609 – 614 (1980)

E. PLATEN

Weak Convergence of Approximations of Itô Integral Equations

In der Arbeit wird die schwache Konvergenz stetig interpolierter Approximationen von Itôschen Integralgleichungen, die in Wagner, Platen [8] vorgestellt wurden, unter der Voraussetzung der Existenz des vierten Momentes des Anfangswertes nachgewiesen.

Weak convergence is proved for continuously interpolated approximations of Itô integral equations, which were presented in Wagner, Platen [8], under the assumption of existence of the 4th moment of the initial value.

В работе доказывается слабая сходимость непрерывно интерполированных приближений к интегральным уравнениям Ито, которые рассматриваются в работе Вагнера, Платена [8], при условии существования четвертого момента начальных данных.

1. Introduction

We are given an ITÔ process on a probability space $[\Omega, \mathfrak{B}, \mathsf{P}]$ by the ITÔ *integral equation*

$$x_t = x_{t_0} + \int_{t_0}^{t} a(u, x_u)\, du + \int_{t_0}^{t} b(u, x_u)\, dw_u \qquad t \in [t_0, T]\,, \tag{1}$$

where $x_t = \{x_t^{(i)}\}_{i=1}^{m}$ and $a(t, x) = \{a^{(i)}(t, x)\}_{i=1}^{m}$ are m-dimensional vectors, $b(t, x) = \{b^{(i,j)}(t, x)\}_{i,j=1}^{m,n}$ is an (m, n)-matrix and $w_t = \{w_t^{(j)}\}_{j=1}^{n}$ is an n-dimensional standard WIENER process $m, n \in \{1, 2, \ldots\}$. $a^{(i)}(t, x)$ and $b^{(i,j)}(t, x)$ are real valued continuous functions on $[t_0, T] \times R^m$ which fulfil the LIPSCHITZ condition

$$\|a(t, x) - a(t, y)\| + \|b(t, x) - b(t, y)\| \leqq K_1 \|x - y\| \; ^{1)}$$

for all $t \in [t_0, T]$ and $x, y \in R^m$. Obviously for all $t \in [t_0, T]$ and $x \in R^m$ the *growth condition* follows

$$\|a(t, x)\| + \|b(t, x)\| \leqq \|a(t, x) - a(t, 0)\| + \|a(t, 0)\| + \|b(t, x) - b(t, 0)\| + \|b(t, 0)\| \leqq$$
$$\leqq 2K_1 \|x\| + \|a(t, 0)\| + \|b(t, 0)\| \leqq K_2(1 + \|x\|)\,.$$

We assume that the initial value x_{t_0} is independent of $\{w_t\}_{t\in[t_0, T]}$ and $\mathsf{E}\,\|x_{t_0}\|^2 < \infty$, then existence and uniqueness of the solutions of (1) are shown in GICHMAN, SKOROCHOD [4] and FRIEDMAN [2].

ITÔ integral equations are of increasing importance in many fields of application. But only a few papers exist which deal with methods for numerical solution of problems connected with ITÔ integral equations. The most effective and universally applicable methods seem to be simulation algorithms which generate step by step approximate realizations.

In MIL'ŠTEIN [6], RAO, BORWANKAR, RAMAKRISHNA [7] and WAGNER, PLATEN [8] such algorithms are considered, where the mean square difference between approximate realizations and the original process or a similar functional converges to zero, if the maximum step size tends to zero. Further GICHMAN, SKOROCHOD [3] (p. 527) and KUSHNER [5] studied weak convergence of similar approximation algorithms where drift and diffusion are assumed to be bounded functions.

In the following we show weak convergence (see BILLINGSLEY [1]) of continuously interpolated versions of the approximations presented in WAGNER, PLATEN [8] under the additional assumption of existence of the 4th moment of the initial value. Boundedness of drift or diffusion is not assumed. Basing on this result it is possible to study functionals of diffusion processes with the help of Monte-Carlo simulation which is also described in [8].

2. Approximation algorithm

We summarize now some results about approximate solutions of (1) contained in WAGNER, PLATEN [8] which will be used in the sequel.

Let $M := \{(j_1, \ldots, j_k): k \in \{1, 2, \ldots\}, j_i \in \{0, 1, \ldots, n\}$ for $i \in \{1, \ldots, k\}\} \cup \{v\}$ denote the set of row vectors $\alpha = (j_1, \ldots, j_k)$ with finite length $l(\alpha) := k$ where $l(v) = 0$. We write $-\alpha$ or $\alpha-$, respectively, if we delete the first or last, resp., component of $\alpha \in M$, $l(\alpha) \geqq 1$. Further we define *recursively stochastic integrals* for $\alpha \in M$, $s, t \in [t_0, T]$, $s \leqq t$,

$$J_\alpha(s, t) := \begin{cases} 1 & \text{for} \quad \alpha = v \\ \int_s^t J_{\alpha-}(s, u)\, dw_u^{(j_k)} & \text{for} \quad \alpha = (j_1, \ldots, j_k) \end{cases} \tag{3}$$

where $dw_u^{(0)} = du$.

Let D denote the set of all continuous functions $f \mid [t_0, T] \times R^m \to R^m$ with continuous partial derivatives of first order $\frac{\partial}{\partial t} f$, $\frac{\partial}{\partial x^{(i)}} f$ and second order $\frac{\partial^2}{\partial x^{(i)}\, \partial x^{(r)}} f$, $i, r \in \{1, \ldots, m\}$. We now introduce the following differential

[1]) $\|\cdot\|$ denotes the norm of a vector or matrix, respectively, which is defined as square root of the sum of its squared components or elements, respectively.

operators defined for $f \in D$:

$$L^{(k)}f := \begin{cases} \frac{\partial}{\partial t} f + \sum_{i=1}^{m} \left(\frac{\partial}{\partial x^{(i)}} f\right) a^{(i)} + \frac{1}{2} \sum_{i,r=1}^{m} \left(\frac{\partial^2}{\partial x^{(i)}\, \partial x^{(r)}} f\right) \sum_{j=1}^{n} b^{(i,j)} b^{(r,j)} & \text{for } k = 0 \\ \sum_{i=1}^{m} \left(\frac{\partial}{\partial x^{(i)}} f\right) b^{(i,k)} & \text{for } k \in \{1, \dots, n\}\,. \end{cases}$$

Further we use the functions $f_\alpha = \{f_\alpha^{(i)}\}_{i=1}^m, \alpha \in M$, where

$$f_\alpha^{(i)} := \begin{cases} 0 & \text{for } \alpha = v \\ a^{(i)} & \text{for } \alpha = (0) \\ b^{(i,j)} & \text{for } \alpha = (j),\ j \in \{1, \dots, n\} \\ L^{(j_1)} f_{-\alpha}^{(i)} & \text{for } \alpha = (j_1, \dots, j_k),\ k \geqq 2\,. \end{cases}$$

Let $n(\alpha)$ denote the number of components of $\alpha \in M$ which are zero. For a given natural number $\varepsilon \in \{1, 2, \dots\}$ we set

$$A_\varepsilon := \{\alpha \in M : l(\alpha) + n(\alpha) \leqq \varepsilon \ \text{ or } \ l(\alpha) = n(\alpha) = (\varepsilon + 1)/2\}\,.$$

We assume for all $\alpha = (j_1, \dots, j_k) \in A_\varepsilon$

$$f_{(j_k)}, f_{(j_{k-1}, j_k)}, \dots, f_{(j_1, \dots, j_k)} \in D$$

and the LIPSCHITZ condition

$$\|f_\alpha(s, x) - f_\alpha(s, y)\| \leqq K_1 \|x - y\|$$

for all $s \in [t_0, T]$ and $x, y \in R^m$. Further we suppose

$$\mathsf{E}\, \|f_\beta(s, x_s)\|^2 \leqq K_3$$

for all $s \in [t_0, T]$ and $\beta \in \{\alpha \in M \setminus A_\varepsilon : -\alpha \in A_\varepsilon\}$. Finally $\Gamma = \{\tau_1, \tau_2, \dots\}$ denotes a set of finite discretizations $\tau = \{t_i\}_{i=0}^N$, $t_0 < t_1 < \dots < t_N = T$, of the interval $[t_0, T]$ and $h(\tau) := \max_{i \in \{0, 1, \dots, N-1\}} (t_{i+1} - t_i)$.

Now the following assertion is proved in [8]:

Theorem 1: *The approximate solution $\{\hat{x}_\tau^\varepsilon(t)\}_{t \in [t_0, T]}$, for a given $\varepsilon \in \{1, 2, \dots\}$ and $\tau \in \Gamma$, defined recursively by*

$$\hat{x}_\tau^\varepsilon(t_0) := x_{t_0}\,, \qquad \hat{x}_\tau^\varepsilon(t) := \hat{x}_\tau^\varepsilon(t_i) + \sum_{\alpha \in A_\varepsilon} f_\alpha(t_i, \hat{x}_\tau^\varepsilon(t_i))\, J_\alpha(t_i, t)\,,$$

$i \in \{0, 1, \dots, N - 1\}$, $t \in [t_i, t_{i+1}]$, possesses the property that

$$\sup_{t \in [t_0, T]} \mathsf{E}\, \|x_t - \hat{x}_\tau^\varepsilon(t)\|^2 \leqq K_4 h(\tau)^\varepsilon\,,$$

where K_4 is a constant not depending on τ.

3. Weak convergence of probability measures

In the following we give some remarks on weak convergence of probability measures. The mentioned facts can be found in detail, for instance, in BILLINGSLEY [1] and other wellknown literature on this subject.

Let $C(t_0, T)$ denote the set of all continuous functions on $[t_0, T]$ with values in R^m. Then a metric exists such that $C(t_0, T)$ is complete and separable. Further $\mathscr{C}$ denotes the σ-algebra of BOREL-measurable sets corresponding to this metric. $\mathscr{M}$ is the family of all probability measures on $[C, \mathscr{C}]$.

Now we suppose a sequence of measures $(Q_q)_{q \in \{1,2,\dots\}}$ and a measure Q are given on $[C, \mathscr{C}]$. We say that the sequence of measures $(Q_q)_{q \in \{1,2,\dots\}}$ *converges weakly* to the measure Q, if for all measurable real valued Q-almost sure continuous and bounded functions f on $C(t_0, T)$

$$\lim_{q \to \infty} \int f(x)\, \mathrm{d}Q_q(x) = \int f(x)\, \mathrm{d}Q(x)\,.$$

It is sufficient for the weak convergence of a sequence of measures on $[C, \mathscr{C}]$ induced by a sequence of processes $(\{y_q(t)\}_{t \in [t_0, T]})_{q \in \{1,2,\dots\}}$ to a measure induced by a process $\{y(t)\}_{t \in [t_0, T]}$ that the finite-dimensional distributions of $\{y_q(t)\}_{t \in [t_0, T]}$ converge to the finite-dimensional distributions of $\{y(t)\}_{t \in [t_0, T]}$ and for all $t \in [t_0, T - \Delta]$, $\Delta \geqq 0$,

$$\sup_{q \in \{1,2,\dots\}} \mathsf{E}\, \|y_q(t + \Delta) - y_q(t)\|^4 \leqq K_5 \Delta^2\,. \tag{6}$$

4. Weak convergence of approximations

For a given $\varepsilon \in \{1, 2, \dots\}$ and $\tau \in \Gamma$ we define the *approximation process $\{x_\tau^\varepsilon(t)\}_{t \in [t_0, T]}$ in the discretization points* analogously to (4) by

$$x_\tau^\varepsilon(t_0) := x_{t_0}\,, \qquad x_\tau^\varepsilon(t_{i+1}) := x_\tau^\varepsilon(t_i) + \sum_{\alpha \in A_\varepsilon} f_\alpha(t_i, x_\tau^\varepsilon(t_i))\, J_\alpha(t_i, t_{i+1})\,, \qquad i \in \{0, 1, \dots, N - 1\}\,. \tag{7}$$

In the remaining time the process $\{x_\tau^\varepsilon(t)\}_{t \in [t_0, T]}$ will be continuously interpolated, such that for all $i \in \{0, 1, \dots, N - 1\}$ and $t, s \in [t_i, t_{i+1}]$, $t \geqq s$,

$$\mathsf{E}\, \|x_\tau^\varepsilon(t) - x_\tau^\varepsilon(s)\|^4 \leqq K_6(1 + \mathsf{E}\, \|x_\tau^\varepsilon(t_i)\|^4)\, (t - s)^2\,. \tag{8}$$

It will be shown in Lemma 4 that this interpolation condition is fulfilled, for instance, for the approximation defined by (4) and usual linear interpolation.

Now we can formulate the following

Theorem 2: *If* $\lim_{q\to\infty} h(\tau_q) = 0$ *and* $\mathsf{E}\|x_{t_0}\|^4 < \infty$ *then for all* $\varepsilon \in \{1, 2, \dots\}$ *the sequence of probability measures on* $[C, \mathscr{C}]$ *induced by the sequence of processes* $(\{x^{\varepsilon}_{\tau_q}(t)\}_{t\in[t_0, T]})_{q\in\{1,2,\dots\}}$ *converges weakly to that induced by the solution of* (1) $\{x_t\}_{t\in[t_0,T]}$.

For the proof of this assertion we study at first some properties of the stochastic integrals defined in (3). We denote by $\mathscr{F}_t$, $t \in [t_0, T]$, the σ-algebra generated by the WIENER process w_s for $s \in [t_0, t]$.

Lemma 1: *For all* $\alpha \in M$ *and* $s, t \in [t_0, T]$, $s \leqq t$, *it is* P*-almost surely that*

$$\mathsf{E}(|J_\alpha(s, t)|^4 \mid \mathscr{F}_s) \leqq 6^{2(l(\alpha)-n(\alpha))}(t-s)^{2(l(\alpha)+n(\alpha))} \tag{9}$$

holds.

Proof: Induction with respect to $l(\alpha)$:

1. $l(\alpha) = 0$: It follows $\alpha = v$. Therefore because of (3) $J_\alpha(s, t) = 1$ and on the other hand $n(\alpha) = 0$. We obtain

$$\mathsf{E}(|J_\alpha(s, t)|^4 \mid \mathscr{F}_s) = 1 = 6^{2(l(\alpha)-n(\alpha))}(t-s)^{2(l(\alpha)+n(\alpha))}.$$

2. $l(\alpha) = k + 1$, $k \geqq 0$, $\alpha = (j_1, \dots, j_{k+1})$:

2.1. $j_{k+1} = 0$: Because of (3) we have

$$J_\alpha(s, t) = \int_s^t J_{\alpha-}(s, u)\, du .$$

Using the CAUCHY-BUNJAKOWSKI inequality we get

$$\mathsf{E}(|J_\alpha(s, t)|^4 \mid \mathscr{F}_s) = \mathsf{E}\left(\left(\left|\int_s^t J_{\alpha-}(s, u)\, du\right|^2\right)^2 \middle| \mathscr{F}_s\right) \leqq \mathsf{E}\left(\left((t-s)\int_s^t |J_{\alpha-}(s, u)|^2\, du\right)^2 \middle| \mathscr{F}_s\right) \leqq$$

$$\leqq (t-s)^2\, \mathsf{E}\left((t-s)\int_s^t |J_{\alpha-}(s, u)|^4\, du \middle| \mathscr{F}_s\right) \leqq (t-s)^3 \int_s^t \mathsf{E}(|J_{\alpha-}(s, u)|^4 \mid \mathscr{F}_s)\, du \leqq$$

$$\leqq (t-s)^3\, 6^{2(l(\alpha-)-n(\alpha-))}(t-s)^{2(l(\alpha-)+n(\alpha-))+1} \leqq 6^{2(l(\alpha)-n(\alpha))}(t-s)^{2(l(\alpha)+n(\alpha))}.$$

2.2. $j_{k+1} \in \{1, \dots, n\}$: Because of (3)

$$J_\alpha(s, t) = \int_s^t J_{\alpha-}(s, u)\, dw_u^{(j_{k+1})} .$$

Using a well-known estimation (see GICHMAN, SKOROCHOD [3] p. 458) we obtain

$$\mathsf{E}(|J_\alpha(s, t)|^4 \mid \mathscr{F}_s) = \mathsf{E}\left(\left|\int_s^t J_{\alpha-}(s, u)\, dw^{(j_{k+1})}\right|^4 \middle| \mathscr{F}_s\right) \leqq 36(t-s)\, \mathsf{E}\left(\int_s^t |J_{\alpha-}(s, u)|^4\, du \middle| \mathscr{F}_s\right) \leqq$$

$$\leqq 6^2(t-s)\int_s^t \mathsf{E}(|J_{\alpha-}(s, u)|^4 \mid \mathscr{F}_s)\, du \leqq 6^2(t-s)\, 6^{2(l(\alpha)-n(\alpha-))}(t-s)^{2(l(\alpha\ )+n(\alpha-))+1} \leqq$$

$$\leqq 6^{2(l(\alpha)-n(\alpha))}(t-s)^{2(l(\alpha)+n(\alpha))}.$$

Lemma 2: *Let* $g(t) = \{g^{(j)}(t)\}_{j=1}^m$ *be an* $\mathscr{F}_t$*-measurable function from* $[t_0, T]$ *into* R^m *and for all discretizations* $\tau \in \Gamma$ *and* $i_1, i_2 \in \{0, 1, \dots, N-1\}$, $i_1 \leqq i_2$,

$$R := \sum_{i=i_1}^{i_2} \mathsf{E}\, \|g(t_i)\|^4\, (t_{i+1} - t_i) < \infty$$

then for all $\alpha \in M \setminus \{v\}$

$$\mathsf{E}\left\|\sum_{i=i_1}^{i_2} g(t_i)\, J_\alpha(t_i, t_{i+1})\right\|^4 \leqq (t_{i_2+1} - t_{i_1})\, R K_7\, 6^{2(l(\alpha)-n(\alpha))}(T - t_0)^{2(l(\alpha)+n(\alpha)-1)} .$$

Proof: Obviously

$$\mathsf{E}\left\|\sum_{i=i_1}^{i_2} g(t_i)\, J_\alpha(t_i, t_{i+1})\right\|^4 \leqq m \sum_{j=1}^m \mathsf{E}\left|\sum_{i=i_1}^{i_2} g^{(j)}(t_i)\, J_\alpha(t_i, t_{i+1})\right|^4 .$$

We set for $j \in \{1, \dots, m\}$, $t \in [t_0, T]$,

$$\hat{g}^{(j)}(t) := \sum_{i=0}^{N-1} g^{(j)}(t_i)\, \chi_{(t_i, t_{i+1}]}(t)$$

and

$$\hat{J}_\alpha(t) = \sum_{i=0}^{N-1} J_\alpha(t_i, t)\, \chi_{(t_i, t_{i+1}]}(t)$$

where

$$\chi_{(t_i, t_{i+1}]}(t) = \begin{cases} 1 & \text{if } t \in (t_i, t_{i+1}] \\ 0 & \text{otherwise} \end{cases},$$

$i \in \{0, 1, \ldots, N-1\}$, then we have for $\alpha = (\alpha-, l)$ because of (3)

$$\sum_{i=i_1}^{i_2} g^{(j)}(t_i)\, J_\alpha(t_i, t_{i+1}) = \int_{t_{i_1}}^{t_{i_2+1}} \hat{g}^{(j)}(t)\, \hat{J}_{\alpha-}(t)\, dw_t^{(l)} .$$

In the case $l = 0$ we get by the Cauchy-Bunjakowski inequality

$$\mathsf{E} \left| \int_{t_{i_1}}^{t_{i_2+1}} \hat{g}^{(j)}(t)\, \hat{J}_{\alpha-}(t)\, dt \right|^4 \leqq (t_{i_2+1} - t_{i_1})^3\, \mathsf{E} \int_{t_{i_1}}^{t_{i_2+1}} |\hat{g}^{(j)}(t)\, \hat{J}_{\alpha-}(t)|^4\, dt .$$

For $l \in \{1, 2, \ldots, n\}$ we obtain with the help of a well-known estimation (see [3], p. 458)

$$\mathsf{E} \left| \int_{t_{i_1}}^{t_{i_2+1}} \hat{g}^{(j)}(t)\, \hat{J}_{\alpha-}(t)\, dw_t^{(l)} \right|^4 \leqq 6^2 (t_{i_2+1} - t)\, \mathsf{E} \int_{t_{i_1}}^{t_{i_2+1}} |\hat{g}^{(j)}(t)\, \hat{J}_{\alpha-}(t)|^4\, dt .$$

On the other hand it follows from Lemma 1

$$\mathsf{E} \int_{t_{i_1}}^{t_{i_2+1}} |\hat{g}^{(j)}(t)\, \hat{J}_{\alpha-}(t)|^4\, dt = \sum_{i=i_1}^{i_2} \mathsf{E}\, |g^{(j)}(t_i)|^4 \int_{t_i}^{t_{i+1}} \mathsf{E} \left(|J_{\alpha-}(t_i, t)|^4 \mid \mathcal{F}_{t_i} \right) dt \leqq$$

$$\leqq \sum_{i=i_1}^{i_2} \mathsf{E}\, |g^{(j)}(t_i)|^4 \int_{t_i}^{t_{i+1}} 6^{2(l(\alpha-) - n(\alpha-))} (t - t_i)^{2(l(\alpha-) + n(\alpha-))}\, dt \leqq$$

$$\leqq 6^{2(l(\alpha-) - n(\alpha-))} (t_{i_2+1} - t_{i_1})^{2(l(\alpha-) + n(\alpha-))}\, R .$$

Combining the assertions above we get directly the desired estimation in Lemma 2.

Now we prove the following property of the approximation processes.

Lemma 3: *If $\mathsf{E}\, ||x_{t_0}||^4 < \infty$ then for all $\varepsilon \in \{1, 2, \ldots\}$, $\tau \in \Gamma$, and $t, s \in [t_0, T]$, $\geqq s$,*

$$\mathsf{E}\, ||x_\tau^\varepsilon(t) - x_\tau^\varepsilon(s)||^4 \leqq K_8 (t - s)^2$$

and

$$\sup_{t \in [t_0, T]} \mathsf{E}\, ||x_\tau^\varepsilon(t)||^4 \leqq K_9$$

where K_8 and K_9 are constants not depending on τ.

Proof: For all $l \in \{0, 1, \ldots, N\}$ we write

$$x_l^\varepsilon := x_\tau^\varepsilon(t_l) ,$$

and set for $i_1 \in \{0, 1, \ldots, N-1\}$ and $i \in \{i_1 + 1, \ldots, N\}$

$$\chi_A(i) := \begin{cases} 1 & \text{if } \max\limits_{l \in \{i_1, \ldots, i\}} \mathsf{E}\, ||x_l^\varepsilon - x_{i_1}^\varepsilon||^4 \leqq A \\ 0 & \text{otherwise} \end{cases}$$

where A is a given positive constant.

Because of (7) and Lemma 2 we get for $i_1 \in \{0, 1, \ldots, N-1\}$ and $i \in \{i_1 + 1, \ldots, N\}$

$$\mathsf{E}\, \chi_A(i)\, ||x_i^\varepsilon - x_{i_1}^\varepsilon||^4 \leqq \mathsf{E}\, \chi_A(i) \left\| \sum_{l=i_1}^{i-1} \sum_{\alpha \in A_\varepsilon} f_\alpha(t, x_l^\varepsilon)\, J_\alpha(t_l, t_{l+1}) \right\|^4 \leqq 8^{\operatorname{card} A_\varepsilon - 1} \sum_{\alpha \in A_\varepsilon} \chi_A(i)\, \mathsf{E} \left\| \sum_{l=i_1}^{i-1} f_\alpha(t_l, x_l^\varepsilon)\, J_\alpha(t_l, t_{l+1}) \right\|^4 \leqq$$

$$\leqq 8^{\operatorname{card} A_\varepsilon - 1} \sum_{\alpha \in A_\varepsilon} \chi_A(i)\, (t_i - t_{i_1})\, K_7 6^{2\varepsilon} (T - t_0)^{2\varepsilon} \sum_{l=i_1}^{i-1} \mathsf{E}\, ||f_\alpha(t_l, x_l^\varepsilon)||^4\, (t_{l+1} - t_l) .$$

For the functions f_α it follows from their continuity and Lipschitz condition a growth condition analogous to (2) and we obtain

$$\mathsf{E}\, \chi_A(i)\, ||x_i^\varepsilon - x_{i_1}^\varepsilon||^4 \leqq c_1 \sum_{\alpha \in A_\varepsilon} \chi_A(i)\, (t_i - t_{i_1}) \sum_{l=i_1}^{i-1} \{ K_1^4 \mathsf{E}\, ||x_l^\varepsilon - x_{i_1}^\varepsilon||^4 + 8 c_2^4 (1 + \mathsf{E}\, ||x_{i_1}^\varepsilon||^4) \}\, (t_{l+1} - t_l) \leqq$$

$$\leqq (t_i - t_{i_1})\, c_3 \sum_{l=i_1}^{i-1} \mathsf{E}\, \chi_A(l)\, ||x_l^\varepsilon - x_{i_1}^\varepsilon||^4\, (t_{l+1} - t_l) + (t_i - t_{i_1})^2 c_4 (1 + \mathsf{E}\, ||x_{i_1}^\varepsilon||^4) .$$

If we set for all $u \in [t_0, T]$

$$i_\tau(u) := \max \{ l \in \{0, 1, \ldots, N-1\} : t_l \leqq t \}$$

and for all $z \in [t_{i_1}, T]$,

$$\alpha(z) := \mathsf{E}\, \chi_A(i_\tau(z))\, ||x_{i_\tau(z)}^\varepsilon - x_{i_1}^\varepsilon||^4$$

then

$$\alpha(z) \leqq (z - t_{i_1})\, c_3 \int_{t_{i_1}}^{z} \alpha(u)\, du + (z - t_{i_1})^2 c_4 (1 + \mathsf{E}\, ||x_{i_1}^\varepsilon||^4) .$$

It follows by a Lemma of Gichman and Skorochod [3] (p. 471)

$$\alpha(z) \leqq (z - t_{i_1})\, c_3 \int_{t_{i_1}}^{z} \exp \{ (z - t_{i_1})\, c_3 (z - u) \}\, (u - t_{i_1})^2 c_4 (1 + \mathsf{E}\, ||x_{i_1}^\varepsilon||^4)\, du + (z - t_{i_1})^2 c_4 (1 + \mathsf{E}\, ||x_{i_1}^\varepsilon||^4) \leqq$$

$$\leqq (z - t_{i_1})^2 c (1 + \mathsf{E}\, ||x_{i_1}^\varepsilon||^4) .$$

Obviously, this estimation is independent of the constant A and it follows for $z = t_i$

$$\mathsf{E}\,||x_i^\varepsilon - x_{i_1}^\varepsilon||^4 \leqq (t_i - t_{i_1})^2\, c_5(1 + \mathsf{E}\,||x_{i_1}^\varepsilon||^4)\,. \tag{10}$$

Using the assumption of the lemma and (7) we obtain for all $l \in \{0, 1, \dots, N\}$

$$\mathsf{E}\,||x_l^\varepsilon||^4 \leqq 8\{\mathsf{E}\,||x_{t_0}||^4 + (T - t_0)^2\, c_5(1 + \mathsf{E}\,||x_{t_0}||^4)\} \leqq c_6\,.$$

Because of (8) it follows for all $t \in [t_0, T]$

$$\mathsf{E}\,||x_\tau^\varepsilon(t)||^4 \leqq 8\{\mathsf{E}\,||x_{i\tau(t)}^\varepsilon||^4 + \mathsf{E}\,||x_\tau^\varepsilon(t) - x_{i\tau(t)}^\varepsilon||^4\} \leqq 8\{c_6 + K_6(1 + c_6)\,(T - t_0)^2\} \leqq K_9\,.$$

Further we get for $t, s \in [t_0, T]$, $t \geqq s$, by (8) and (10)

$$\mathsf{E}\,||x_\tau^\varepsilon(t) - x_\tau^\varepsilon(s)||^4 \leqq 64\{\mathsf{E}\,||x_\tau^\varepsilon(t) - x_{i\tau(t)}^\varepsilon||^4 + \mathsf{E}\,||x_{i\tau(t)}^\varepsilon - x_{i\tau(s)+1}^\varepsilon||^4 + \mathsf{E}\,||x_{i\tau(s)+1}^\varepsilon - x_\tau^\varepsilon(s)||^4\} \leqq$$
$$\leqq 64\{K_6(1 + c_6)\,(t - t_{i\tau(t)})^2 + c_5(1 + c_6)\,(t_{i\tau(t)} - t_{i\tau(s)+1})^2 + K_6(1 + c_6)\,(t_{i\tau(s)+1} - s)^2\} \leqq K_8(t - s)^2\,.$$

Now we arrive at the

Proof of Theorem 2:

1. At first we show that finite-dimensional distributions of $\{x_\tau^\varepsilon(t)\}_{t \in [t_0, T]}$, $\tau \in \Gamma$, converge for $h(\tau) \to 0$ to finite-dimensional distributions of $\{x_t\}_{t \in [t_0, T]}$.

Because of Theorem 1 and (7) we have for all $i \in \{0, 1, \dots, N\}$

$$\mathsf{E}\,||x_i^\varepsilon - x_{t_i}||^2 \leqq K_4 h(\tau)^\varepsilon \leqq c_1 h(\tau)\,.$$

Further we have for all $t, s \in [t_0, T]$, $t \geqq s$, the estimation (see GICHMAN, SKOROCHOD [4] and FRIEDMAN [2])

$$\mathsf{E}\,||x_t - x_s||^2 \leqq c_2(t - s)\,.$$

Using our interpolation condition (8) and Lemma 3 we obtain for all $i \in \{0, 1, \dots, N - 1\}$ and $t \in [t_i, t_{i+1}]$ by the LJAPUNOV inequality

$$\mathsf{E}\,||x_\tau^\varepsilon(t) - x_i^\varepsilon||^2 \leqq (\mathsf{E}\,||x_\tau^\varepsilon(t) - x_i^\varepsilon||^4)^{2/4} \leqq (K_6(1 + K_9)\,h(\tau)^2)^{1/2} \leqq c_3 h(\tau)\,,$$

and therefore

$$\mathsf{E}\,||x_\tau^\varepsilon(t) - x_t||^2 \leqq 3\{\mathsf{E}\,||x_\tau^\varepsilon(t) - x_i^\varepsilon||^2 + \mathsf{E}\,||x_i^\varepsilon - x_{t_i}|| + \mathsf{E}\,||x_{t_i} - x_t||^2\} \leqq$$
$$\leqq 3\{c_3 h(\tau) + c_1 h(\tau) + c_2 h(\tau)\} \leqq c_4 h(\tau)\,.$$

For the convergence of the finite-dimensional distributions we have to show that the corresponding characteristic functions converge. Let $s_1, s_2, \dots, s_\nu \in [t_0, T]$ be given, then for any $\lambda_1, \lambda_2, \dots, \lambda_\nu \in R^m$ we have

$$\left|\mathsf{E}\exp\left\{i\sum_{l=1}^{\nu}(\lambda_l, x_\tau^\varepsilon(s_l))\right\} - \mathsf{E}\exp\left\{i\sum_{l=1}^{\nu}(\lambda_l, x_{s_l})\right\}\right| \leqq \mathsf{E}\left|\exp\left\{i\sum_{l=1}^{\nu}(\lambda_l, x_\tau^\varepsilon(s_l) - x_{s_l})\right\} - 1\right| \leqq^{2)}$$
$$\leqq \mathsf{E}\left|\sum_{l=1}^{\nu}(\lambda_l, x_\tau^\varepsilon(s_l) - x_{s_l})\right| \leqq \left(\sum_{l=1}^{\nu}||\lambda_l||^2 \sum_{l=1}^{\nu}\mathsf{E}\,||x_\tau^\varepsilon(s_l) - x_{s_l}||^2\right)^{1/2} m \leqq (c_5\nu c_4 h(\tau))^{1/2}\, m \leqq c_6 h(\tau)^{1/2}\,.$$

It follows, that for any $\lambda_1, \lambda_2, \dots, \lambda_\nu \in R^m$ the characteristic functions converge for $h(\tau) \to 0$.

2. As remarked in the third section, it remains to show condition (6) that is for all $t \in [t_0, T - \Delta]$, $\Delta \geqq 0$,

$$\sup_{\tau \in \Gamma}\mathsf{E}\,||x_\tau^\varepsilon(t + \Delta) - x_\tau^\varepsilon(t)||^4 \leqq K_5\Delta^2\,,$$

where K_5 is any constant not depending on t, Δ or τ. But because of Lemma 3 this condition is obviously fulfilled and the proof of Theorem 2 is complete.

We prove now as a comment on the interpolation condition (8)

Lemma 4: *For the approximation process defined by* (4) *and also for usual linear interpolation the interpolation condition* (8) *is fulfilled.*

Proof:

1. Approximation defined by (4): From (4) we get for all $i \in \{0, 1, \dots, N - 1\}$ and $t, s \in [t_i, t_{i+1}]$, $t \geqq s$,

$$x_\tau^\varepsilon(t) - x_\tau^\varepsilon(s) = \sum_{\alpha \in A_\varepsilon} f_\alpha(t_i, x_i^\varepsilon)\,(J_\alpha(t_i, t) - J_\alpha(t_i, s))\,.$$

For the functions f_α we have already remarked that a growth condition is valid such that

$$\mathsf{E}\,||x_\tau^\varepsilon(t) - x_\tau^\varepsilon(s)||^4 \leqq 8^{\operatorname{card} A_\varepsilon - 1}\sum_{\alpha \in A_\varepsilon}\mathsf{E}\,|c_1(1 + ||x_i^\varepsilon||)|^4\,|J_\alpha(t_i, t) - J_\alpha(t_i, s)|^4 \leqq$$
$$\leqq c_2\sum_{\alpha \in A_\varepsilon}\mathsf{E}(1 + ||x_i^\varepsilon||^4)\,\mathsf{E}(|J_\alpha(t_i, t) - J_\alpha(t_i, s)|^4 \mid \mathcal{F}_{t_i})\,.$$

On the other hand it follows from (3) for $\alpha = (j_1, j_2, \dots, j_k) \in A_\varepsilon$ with $j_k = 0$

$$\mathsf{E}(|J_\alpha(t_i, t) - J_\alpha(t_i, s)|^4 \mid \mathcal{F}_{t_i}) = \mathsf{E}\left(\left|\int_s^t J_{\alpha-}(t_i, u)\,du\right|^4 \,\middle|\, \mathcal{F}_{t_i}\right) \leqq (t - s)^2\,\mathsf{E}\left(\left|\int_s^t |J_{\alpha-}(t_i, u)|^2\,du\right|^2 \,\middle|\, \mathcal{F}_{t_i}\right) \leqq$$
$$\leqq (t - s)\,c_3\int_s^t \mathsf{E}(|J_{\alpha-}(t_i, u)|^4 \mid \mathcal{F}_{t_i})\,du$$

2) $(\cdot, \cdot)$ denotes the usual scalar product.

and with $j_k \in \{1, 2, \dots, n\}$ (see GICHMAN, SKOROCHOD [3], p. 458)

$$\mathsf{E}\left(|J_\alpha(t_i, t) - J_\alpha(t_i, s)|^4 \mid \mathcal{F}_{t_i}\right) = \mathsf{E}\left(\left|\int_s^t J_{\alpha-}(t_i, u)\, dw_u^{(j_k)}\right|^4 \middle| \mathcal{F}_{t_i}\right) \leqq (t-s)\, 36 \int_s^t \mathsf{E}\left(|J_{\alpha-}(t_i, u)|^4 \mid \mathcal{F}_{t_i}\right) du .$$

Therefore we get by Lemma 1

$$\mathsf{E}\,||x_\tau^\varepsilon(t) - x_\tau^\varepsilon(s)||^4 \leqq c_4(1 + \mathsf{E}\,||x_i^\varepsilon||^4)\,(t-s) \int_s^t 6^{2\varepsilon} \max\left(1, (T-t_0)\right)^{2(\varepsilon+1)} du \leqq K_6(1 + \mathsf{E}\,||x_i^\varepsilon||^4)\,(t-s)^2 .$$

2. Linear interpolated approximation:
Obviously we have for all $i \in \{0, 1, \dots, N-1\}$ and $t, s \in [t_i, t_{i+1}]$, $t \geqq s$,

$$x_\tau^\varepsilon(t) - x_\tau^\varepsilon(s) = x_\tau^\varepsilon(t_i) + \frac{x_\tau^\varepsilon(t_{i+1}) - x_\tau^\varepsilon(t_i)}{t_{i+1} - t_i}(t - t_i) - \left(x_\tau^\varepsilon(t_i) + \frac{x_\tau^\varepsilon(t_{i+1}) - x_\tau^\varepsilon(t_i)}{t_{i+1} - t_i}(s - t_i)\right)$$

$$= \frac{x_\tau^\varepsilon(t_{i+1}) - x_\tau^\varepsilon(t_i)}{t_{i+1} - t_i}(t - s)$$

Substituting (7) we obtain

$$\mathsf{E}\,||x_\tau^\varepsilon(t) - x_\tau^\varepsilon(s)||^4 = \mathsf{E}\,\Big|\Big|\sum_{\alpha \in A_\varepsilon} f_\alpha(t_i, x_i^\varepsilon)\, J_\alpha(t_i, t_{i+1})\Big|\Big|^4 \frac{(t-s)^4}{(t_{i+1} - t_i)^4} \leqq$$

$$\leqq c_2 \frac{(t-s)^2}{(t_{i+1} - t_i)^2} \sum_{\alpha \in A_\varepsilon} \mathsf{E}\,(1 + ||x_i^\varepsilon||^4)\, \mathsf{E}\left(|J_\alpha(t_i, t_{i+1})|^4 \mid \mathcal{F}_{t_i}\right).$$

Because of Lemma 1 it follows

$$\mathsf{E}\,||x_\tau^\varepsilon(t) - x_\tau^\varepsilon(s)||^4 \leqq c_3 \frac{(t-s)^2}{(t_{i+1} - t_i)^2}(1 + \mathsf{E}\,||x_i^\varepsilon||^4)\, 6^{2\varepsilon}(t_{i+1} - t_i)^2 \max\left(1, (t_{i+1} - t_i)\right)^{2\varepsilon} \leqq$$

$$\leqq K_6(1 + \mathsf{E}\,||x_i^\varepsilon||^4)\,(t-s)^2 .$$

References

1 BILLINGSLEY, P., Convergence of Probability Measures, Wiley, New York 1968.
2 FRIEDMAN, A., Stochastic Differential Equations and Applications. I, II, New York 1975.
3 Гихман, И. И.; Скороход, А. В., Введение в теорию случайных процессов, Москва 1977.
4 Гихман, И. И.; Скороход, А. В., Стохастические дифференциальные уравнения, Киев 1968.
5 KUSHNER, H. J., On the weak convergence of interpolated MARKOV chains to a diffusion, Ann. Probability 2 (1974), 40—50.
6 Мильштейн, Г. Н., Приближенное интегрирование стохастических дифференциальных уравнений. Теория вероятн. и её применения (1974) 3, 583—588.
7 RAO, N. J.; BORWANKAR, J. D.; RAMAKRISHNA, D., Numerical solution of ITÔ integral equations, SIAM J. Control 12 (1974) 1, 124—139.
8 WAGNER, W.; PLATEN, E., Approximation of ITÔ integral equations. Preprint ZIMM, AdW der DDR, Berlin 1978.

Eingereicht: 7. 4. 1979

Anschrift: Dr. ECKHARD PLATEN, ZIMM der AdW der DDR, Mohrenstraße 39, DDR-1080 Berlin

ZAMM 60, 615 – 622 (1980)

P. DE MOTTONI / A. TESEI

A Comparative Analysis of Some Semi-Linear Parabolic Systems Modelling the Dynamics of Fast Nuclear Reactors

Von einer Klasse parabolischer Systeme ausgehend, welche das Verhalten eines Kernreaktors beschreibt, werden zwei Arten modifizierter physikalisch bedeutsamer Systeme untersucht. Eine wesentliche Rolle wird dabei von der Geschwindigkeit der Neutronen $1/\varepsilon$ gespielt. Es wird gezeigt, daß, während solche Systeme bei festem $\varepsilon > 0$ dem ursprünglichen System äquivalent sind, sie sich jedoch, beim Grenzübergang $\varepsilon \to 0$, in einer durchaus verschiedenen Weise verhalten. Insbesondere wird das Auftreten von ungedämpften Schwingungen, welches das ursprüngliche System kennzeichnet, vermieden.

Starting from a class of parabolic systems considered in the literature for describing the operation of fast nuclear reactors, two kinds of modified systems are investigated: one accounts for retarded neutrons, the other one involves a small neutr on feedback. It is shown that these systems, while equivalent to the original ones for any finite value of the neutron velocity, behave in a substantially different way in the limiting case of diverging neutron velocity. In particular, considering reta d ed neutrons or neutron feedback terms avoids the appearance of undamped oscillations, which characterizes the behavior of the original systems in the same limiting case.

Исходя из класса параболических систем, описывающего поведение ядерного реактора, рассматриваются две модифицированных, физически важных системы: Одна учитывает замедленные нейтроны, другая содержит обратную связь нейтронов. При этом существенную роль играет скорость $1/\varepsilon$ нейтронов. Показывается, что эти системы, будучи эквивалентными исходной системе при фиксированном $\varepsilon > 0$, имеют совершенно различное поведение при $\varepsilon \to 0$. В частности, нет появления незатухающих колебаний, характерного для исходной системы.

1. Introduction

In past years several investigations of nuclear reactor dynamics focused on a family of systems of parabolic equations, describing the time behavior of both the (fast) neutron density and one characteristic temperature [1, 9, 10, 12]. This family of models, though ignoring delayed neutrons and considering only one "energy group" [9], is generally believed to provide a qualitatively faithful representation of relevant phenomena — accidents in the operation of the reactor among others. Such equations, after a suitable redefinition of the coefficients, can be written as follows:

$$\begin{aligned} \varepsilon\, \partial_t u &= \Delta u + \xi u - uv \\ \partial_t v &= u - v \end{aligned} \qquad \text{in} \quad (0, \infty) \times \Omega\,, \tag{1.1}$$

where u denotes the neutron density, v the temperature, $\Omega \subset \mathbb{R}^n$ ($n \leqq 3$) an open bounded set, representing the region where the reaction takes place; ε and ξ are positive parameters, describing respectively the inverse of the neutron velocity and the combined effect of neutron self-multiplication and absorption. Equation (1.1) should be supplemented with initial and boundary conditions in order to define a problem which is both mathematically well-posed and physically reasonable: as a model for possibly more complicated situations, we shall work with DIRICHLET homogeneous boundary conditions.

In [7] the (unique) nontrivial nonnegative equilibrium of (1.1), corresponding to the operating point of the reactor in the super-critical case, was shown to be asymptotically stable (in an appropriate norm, having a physically satisfactory interpretation). On the other hand, the high velocity of neutrons (i. e., $\varepsilon \ll 1$) motivated the singular perturbation analysis carried out in [8], which in particular proves the existence of undamped oscillations in the limiting case $\varepsilon \to 0$. Now, undamped oscillations being dangerous from the point of view of reactor safety, it is important to have as detailed informations as possible on their actual occurrence. As we shall see in the present note, they seem to be a peculiarity of system (1.1), which disappears when the effect of retarded neutrons is taken into account, or an arbitrarily small neutron feedback is added to the right-hand side of the first equation in (1.1).

To be specific, we shall be concerned with the following system

$$\begin{aligned} \varepsilon\, \partial_t u &= \Delta u + \xi u - uv + \sigma w \\ \partial_t v &= u - v \\ \partial_t w &= u - \beta w \end{aligned} \qquad \text{in} \quad (0, \infty) \times \Omega\,, \tag{1.2}$$

where w denotes the density of retarded neutrons, β, σ are positive parameters, and the other quantities are as above. We shall also consider the system

$$\begin{aligned} \varepsilon\, \partial_t u &= \Delta u + \xi u - (\gamma u + |u|^\alpha u)\, v \\ \partial_t v &= u - v \end{aligned} \qquad \text{in} \quad (0, \infty) \times \Omega\,, \tag{1.3}$$

α, γ being positive parameters. Both systems (1.2) and (1.3) should be supplemented with boundary and initial conditions as specified below.

The main result of our investigation is that systems (1.2), (1.3), though sharing with (1.1) both stiffness [13] and the qualitative behavior for any positive value of ε, do behave quite differently in the limit $\varepsilon \to 0$. In fact, we shall see that both (1.2) and (1.3) admit a unique nontrivial equilibrium solution (for any $\beta, \sigma, \alpha, \gamma, \varepsilon > 0$ and ξ larger than a critical value ξ_0) ,whose components are nonnegative, and which is asymptotically stable in the supremum norm. On the other hand, by singular perturbation techniques, we shall show that the solutions of such systems can be represented as sums of an inner and an outer expansions: then in the lowest order, solutions are seen to evolve towards the asymptotically stable equilibrium solution — a behavior which is completely different from that of the lowest order terms in the solutions of (1.1).

As a heuristic explanation of such different behaviors, let us consider the ordinary differential systems we formally obtain from (1.1) to (1.3) by dropping the LAPLACians:

$$\begin{aligned} \varepsilon u' &= \xi u - uv \\ v' &= u - v, \end{aligned} \qquad t \in (0, \infty), \qquad ' = \mathrm{d}/\mathrm{d}t; \tag{1.4}$$

$$\begin{aligned} \varepsilon u' &= \xi u - uv + \sigma w \\ v' &= u - v \\ w' &= u - \beta w, \end{aligned} \tag{1.5}$$

$$\begin{aligned} \varepsilon u' &= \xi u - v(\gamma u + |u|^\alpha u) \\ v' &= u - v. \end{aligned} \tag{1.6}$$

It is easily proved that each of the systems above has a unique equilibrium solution in the positive orthant, which is asymptotically stable for any positive value of the parameters ($\gamma = 0$ is allowed as well). A closer investigation shows, in fact that the linearization of the right-hand sides of the above systems on the respective positive equilibrium has, for small ε, the following eigenvalues:

In case of (1.4)

$$(1/2)\left(-1 \pm ((4 - \varepsilon)/\varepsilon)^{1/2}\right).$$

In case of (1.5)

$$-1, \qquad -(\sigma + 1)/\sigma + O(\varepsilon), \qquad -\sigma/\varepsilon + O(1).$$

In case of (1.6)

$$-(\alpha + 1)/\alpha + O(\varepsilon), \qquad -\xi\alpha/\varepsilon + O(1)$$

(for definiteness, we have put $\beta = 1$ in (1.5) and $\gamma = 0$ in (1.6)).

Thus, while in the first case the divergence of the imaginary part of the eigenvalues as $\varepsilon \to 0$ suggests that very fast oscillations with zero damping should occur, in the remaining cases only one real root diverges, going to $-\infty$, the other ones being fixed or converging to a finite strictly negative limit. Hence the latter systems, though genuinely stiff for small ε, are much less wild than the original one.

To give some further insight into the spirit of the singular perturbation approach we are going to fo'low, consider a general ordinary differential system of the form

$$\varepsilon u' = f(u, v), \qquad v' = g(u, v), \tag{1.7}$$

f, g being smooth functions. Suppose there is an equilibrium solution $(\hat{u}, \hat{v})$: $f(\hat{u}, \hat{v}) = g(\hat{u}, \hat{v}) = 0$. As we are interested in the small ε-behavior, we shall represent u, v as sums of two asymptotic power series in ε, one in the slow variable t the other one in the fast variable t/ε

$$u = \sum \varepsilon^k \bar{u}_k(t) + \sum \varepsilon^k \tilde{u}_k(t/\varepsilon),$$

and similarly for v. The large time behavior is described by the overbarred quantities (outer solution), which in the lowest order of approximation obey the following system (referred to as "reduced system" or system describing the motion on the "slow manifold")

$$\begin{aligned} 0 &= f(u, v) \\ v' &= g(u, v) \end{aligned} \tag{1.8}$$

(where only the initial datum for v is to be assigned). Then it is easy to prove (see [14, 15] for related results):

Proposition 1.1: *If $(\hat{u}, \hat{v})$ is linearly asymptotically stable for* (1.7) *and, in addition, $f_u(\hat{u}, \hat{v}) < 0$, then it is asymptotically stable for* (1.8) *as well.*

Roughly speaking, this means that linear asymptotic stability of $(\hat{u}, \hat{v})$ implies, provided $f_u(\hat{u}, \hat{v}) < 0$, that asymptotic stability of $(\hat{u}, \hat{v})$ holds with respect to the motion on the slow manifold.

Then one way of expressing our results consists in regarding them as instances of a partial differential equation counterpart of the above proposition (see Theorem 4.3 below in particular).

For the sake of completeness, let us briefly sketch the proof of Proposition 1.1: local existence and uniqueness of the solution of (1.8) follow because of the implicit function theorem, as $f_u(\hat{u}, \hat{v}) < 0$, whenever the initial datum for v is taken sufficiently close to $\hat{v}$. Then linearize the second equation about $\hat{v}$: writing $\hat{f}_u$ for $f_u(\hat{u}, \hat{v})$, etc, we get

$$\psi' = (\hat{g}_u(\mathrm{d}u/\mathrm{d}v)_{\hat{v}} + \hat{g}_v)\,\psi = (-\hat{g}_u\hat{f}_u^{-1}\hat{f}_v + \hat{g}_v)\,\psi.$$

Now, the expression in brackets in the last formula is nothing but $\hat{f}_u^{-1}$ times the JACOBIan determinant of f, g at (u, v): as the latter, because of the linear stability hypothesis, is positive, linear stability for (1.8) holds. That linear stability implies (full) asymptotic stability is a matter of standard arguments.

The paper is organized as follows: in Sect. 2 we shall introduce the models and state the main results; in Sect. 3 we prove the existence theorems for the solutions and the asymptotic stability results for the nontrivial equilibria; Sect. 4 is devoted to the derivation of asymptotic expansions and to the investigation of the large time behavior of the lowest order components.

2. Presentation of the models and statement of results

Let Ω be an open bounded set in $\mathbb{R}^n$, representing the space domain where the reaction takes place (hence we shall be interested in $n = 1, 2, 3$), with boundary $\partial\Omega$. We shall define $C(\bar{\Omega})$ as the BANACH space of all continuous functions on $\bar{\Omega}$, endowed with the usual uniform norm: $u \to |u| = \sup \{|u(x)|; x \in \Omega\}$. We shall also consider $C_0(\bar{\Omega})$, the (BANACH) subspace of $C(\bar{\Omega})$, consisting of all functions vanishing on $\partial\Omega$. By $C^+(\bar{\Omega})$, $C_0^+(\bar{\Omega})$, respectively, we shall denote the cones in $C(\bar{\Omega})$, $C_0(\bar{\Omega})$ respectively, of the nonnegative functions on $\bar{\Omega}$. We shall also work with the SOBOLEV spaces $H^p = W^{2,p}(\Omega)$, and we define $H_0^p = H_0^1 \cap H_p$ for $p \geqq 1$. The spaces C_0 and H_0^p are useful to deal with DIRICHLET homogeneous boundary conditions, we are mainly interested in. The principal eigenvalue of the LAPLACian Δ with such boundary conditions will be denoted by ξ_0, $(\xi_0 > 0)$.

In the following, we shall consider *equations for the neutron densitiy* $u = u(t, x)$ *and for the temperature* $v = v(t, x)$; $t \in [0, \infty)$, $x \in \bar{\Omega}$. The *original model* [1, 7, 8, 9, 10, 11] reads, after renormalizing the constants

$$\begin{aligned} \varepsilon\, \partial_t u &= \Delta u + \xi u - uv \\ \partial_t v &= u - v \end{aligned} \quad \text{in} \quad (0, \infty) \times \Omega; \tag{2.1}$$

$$\begin{aligned} u &= 0 && \text{in} \quad (0, \infty) \times \partial\Omega \\ u &= U, \quad v = V && \text{in} \quad \{0\} \times \Omega . \end{aligned} \tag{2.2}$$

where $\varepsilon > 0$, $\xi > \xi_0$, $V \in C^+(\bar{\Omega})$, $U \in C_0^+(\bar{\Omega})$.

The *model taking retarded neutrons into account* is

$$\begin{aligned} \varepsilon\, \partial_t u &= \Delta u + \xi u - uv + \sigma w \\ \partial_t v &= u - v \qquad\qquad \text{in} \quad (0, \infty) \times \Omega . \\ \partial_t w &= u - \beta w . \end{aligned} \tag{2.3}$$

Here $w = w(t, x)$ denotes the density of delayed neutrons, β and σ are positive parameters; the parameters ξ, ε and the boundary conditions are as in (2.2). The *initial conditions* are

$$u = U, \qquad v = V, \qquad w = W \qquad \text{in} \quad \{0\} \times \Omega , \tag{2.4}$$

where $U \in C_0^+(\bar{\Omega})$; $V, W \in C^+(\bar{\Omega})$.

Our *second modified model* reads

$$\begin{aligned} \varepsilon\, \partial_t u &= \Delta u + \xi u - v(\gamma u + |u|^\alpha u) \\ \partial_t v &= u - v , \end{aligned} \tag{2.5}$$

where $\alpha > 0$, $\gamma \geqq 0$, ε, ξ and boundary conditions are as in (2.2); initial conditions are $u = U$, $v = V$ in $\{0\} \times \Omega$.

As anticipated in Sect. 1, (2.3) describes the effect of retarded neutrons, too, along the lines laid down in Ref. [9]. As to (2.5), the term $\gamma u + |u|^\alpha u$, replacing u in (2.1), is a perturbation of the feedback allowing for a stronger u-dependence.

It should be observed that, in any of the models considered above, the quantity v in the feedback term could be replaced by an arbitrary smooth convex function $g(v)$, without altering the qualitative features of the models, nor their limiting behavior as ε goes to zero. Further generalizations are also possible, which do not lead to substantial changes in our treatment: e.g.,

(i), replacing Δ by an elliptic differential operator of the second order, with possibly discontinuous coefficients [7];

(ii) allowing for an x-dependence in the coefficients [7];

(iii) using boundary conditions of mixed type: $(\partial u/\partial \nu) + hu = 0$ on $\partial\Omega$.

The results can be synthesized as follows:

A. *For* $\xi > \xi_0$, *and any value of the parameters* $\varepsilon, \sigma, \beta, \alpha > 0, \gamma \geqq 0$, *the system* (2.3) ((2.5), *respectively*) *possesses a unique equilibrium solution* $(\hat{u}, \hat{v}, \hat{w})$ $((\hat{u}, \hat{v})$, *respectively*), *consisting of smooth functions, vanishing on* $\partial\Omega$, *and positive inside* Ω. *This equilibrium solution is asymptotically stable in the uniform norm.*

B. *For any* $p \geqq 1$ *and any* (V, W) *in a suitable* (ε-*independent*) $H^p \oplus H^p$ *neighbourhood of* $(\hat{v}, \hat{w})$, *the system* (2.3) *has a solution whose* ε-*power series expansion*

$$\begin{aligned} u(t, x) &= \hat{u}(x) + \sum \varepsilon^k \bar{\varphi}_k(t, x) + \sum \varepsilon^k \tilde{\varphi}_k(t/\varepsilon, x) , \\ v(t, x) &= \hat{v}(x) + \quad {}^k\tilde{\psi}_k(t, x) + \sum \varepsilon^k \bar{\psi}_k(t/\varepsilon, x) , \\ w(t, x) &= \hat{w}(x) + \sum \varepsilon^{k-}{}_k(t, x) + \sum \varepsilon^k \tilde{\chi}_k(t/\varepsilon, x) , \end{aligned} \tag{2.6}$$

satisfies, at the lowest order, the following properties

i) $\tilde{\psi}_0(t/\varepsilon, x) = 0$, $\bar{\chi}_0(t/\varepsilon, x) = 0$ *for alle* t, x, ε.

ii) $\tilde{\varphi}_0(t/\varepsilon, x)$ *decays exponentially to zero in the fast time scale* t/ε, *in the Sobolev space* H_0^{p+2}.

iii) $\bar{\psi}_0(t, x)$, $\bar{\chi}_0(t, x)$ *converge to zero in* H^p, *while* $\bar{\varphi}_0(t, x)$ *converges to zero in* H_0^{p+2}.

Similar properties hold true for the solution of system (2.5).

Remarks:

1. The twilded quantities in (2.6), defined in terms of the fast time variable $t/\varepsilon = \tau$ take care of the initial transient (inner expansion); the overbarred ones describe on the other hand the large time behavior (outer expan-

sion). The result B. can be therefore resumed by saying that in the lowest order of approximation the deviation from the equilibrium of the u-component splits into a transient term and in a long-range one, converging, according to the respective time scales, to zero. No transient terms are instead present for the deviations relative to the remaining components, that simply converge to zero in the slow time scale.

2. In contrast, it was proved in [8] that for (2.1) different expansions hold, whose lowest order term does not converge to the equilibrium, but exhibits undamped oscillations.

3. The convergence results presented above could be easily extended to the next orders of approximation.

3. Existence and stability results for fixed ε.

Problems (2.3), (2.5) can be formulated as evolution equations in a convenient BANACH space $\mathcal{B}$. In case of (2.3), we take $\mathcal{B} = C_0(\bar{\Omega}) \oplus C(\bar{\Omega}) \oplus C(\bar{\Omega})$, while for (2.5) $\mathcal{B} = C_0(\bar{\Omega}) \oplus C(\bar{\Omega})$. These spaces are endowed with the norm given by the sum of the supremum norms of each component. Both problems can be then written as an autonomous equation in the corresponding BANACH space $\mathcal{B}$:

$$\begin{aligned} &\frac{dx}{dt} = Ax + f(x) \qquad (t > 0)\,, \\ &x(0) = X \qquad\qquad (X \in \mathcal{B})\,. \end{aligned} \tag{3.1}$$

Here A represents the linear part of the right hand side of (2.3) or (2.5); nonlinear terms are accounted for by f. The local existence and uniqueness for the solution of (3.1) can be proved in a standard way, using the following facts:

i) A generates an analytic semigroup $G(t)$ on $\mathcal{B}$;

ii) f is locally LIPSCHITZ continuous.

As to the global existence, it is ensured by the following properties:

iii) $G(t)$ and the resolvent of f leave the nonnegative cone $\mathcal{B}^+$ of $\mathcal{B}$[1]) invariant;

iv) nonnegative solutions satisfy on any interval $[0, T]$ an a-priori estimate. For details, the reader is referred to [7, 8].

Let us now turn to the equilibrium solutions. *Equilibrium solutions* of (2.3) are characterized as the solutions of

$$\begin{aligned} &0 = \Delta u + \xi u - uv + \sigma w \\ &0 = u - v \qquad\qquad \text{in } \Omega \\ &0 = u - \beta w \\ &u = 0 \qquad\qquad \text{in } \partial\Omega\,. \end{aligned} \tag{3.2}$$

This leads to the single equation for u

$$\Delta u + \left(\xi + \frac{\sigma}{\beta}\right) u - u^2 = 0 \quad \text{in } \Omega\,, \qquad u = 0 \quad \text{in } \partial\Omega\,, \tag{3.3}$$

whose solutions determine the v- and w-components via the second and third equations of (3.2). As well known, (3.3) has a unique nontrivial solution, which is positive inside Ω. We shall denote it by $\hat{u}$; accordingly, we put $\hat{v} = \hat{u}$, $\hat{w} = (1/\beta)\,\hat{u}$. We have $\hat{u} \in C^\infty(\Omega) \cap C_0^+(\bar{\Omega})$, and, because of the maximum principle:

$$\sup \hat{u} \leqq \xi + (\sigma/\beta)\,. \tag{3.4}$$

In addition, as the (second-order, elliptic) operator defined by $\Delta + \left(\xi + \frac{\sigma}{\beta}\right) - \hat{u}$ with DIRICHLET homogeneous boundary conditions has zero as eigenvalue corresponding to the positive eigenfunction $\hat{u}$, it follows (see, e.g. [6])

$$\left\langle\left(\Delta + \left(\xi + \frac{\sigma}{\beta}\right) - \hat{u}\right)\varphi, \varphi\right\rangle \leqq 0, \ \varphi \in H_0^1 \tag{3.5}$$

where $\langle\cdot,\cdot\rangle$ denotes the scalar product in $L^2(\Omega)$. In other words, the spectrum of $\Delta + \left(\xi + \frac{\sigma}{\beta}\right) - \hat{u}$ lies on the nonpositive real axis. Moreover, it can be shown that this operator generates an analytic semigroup of nonpositive type, both in $L^2(\Omega)$ and in $C_0(\bar{\Omega})$.

We shall now sketch the proof of the asymptotic stability of the equilibrium solution col $(\hat{u}, \hat{u}, (1/\beta)\,\hat{u})$ in the space $\mathcal{B}$. To this end, we use the same linearization argument as in [7].

The basic object is is the (FRÉCHET) derivative of the right-hand side of (2.3), namely the operator valued matrix

$$\mathfrak{A} = \begin{pmatrix} \frac{1}{\varepsilon}(\Delta + \xi - \hat{u}) & -\hat{u}/\varepsilon & \sigma/\varepsilon \\ 1 & -1 & 0 \\ 1 & 0 & -\beta \end{pmatrix}.$$

[1]) Consisting of triples, respectively couples, of nonnegative functions.

An important point is now that $\mathfrak{A}$ generates a strongly continuous (as a matter of fact, analytic) semigroup on $\mathcal{B}$ of strictly negative type. To see this, one needs to prove

i) that $(\mathfrak{A} - \mu)^{-1}$ is well-defined in a sector
$\Sigma(\mu_0, \theta) = \{\mu \in \mathrm{C};\ |\arg(\mu + \mu_0)| < \pi/2 + \theta\}$, $(\mu_0 > 0, \theta \in (0, \pi/2))$;
ii) that the operator norm of $(\mathfrak{A} - \mu)^{-1}$ verifies
$\|(\mathfrak{A} - \mu)^{-1}\| < C/|\mu + \mu_0|$,
C being a positive constant, for any $\mu \in \Sigma(\mu_0, \theta)$ (for some $\mu_0 > 0$, $\theta \in (0, \pi/2)$).

We outline here the proof of the first step: from the resolvent equation for $\mathfrak{A}$:

$$\begin{aligned}&(1/\varepsilon)(\Delta + \xi - \hat{u})\varphi - \mu\varphi - (\hat{u}/\varepsilon)\psi + (\sigma/\beta)\chi = u\\ &\varphi - (1 + \mu)\psi = v\\ &\varphi - (\beta + \mu)\chi = w\end{aligned} \tag{3.6}$$

we obtain, upon appropriate substitutions, the equation

$$A(\mu)\varphi = (1/\varepsilon)\big(\sigma w/(\beta + \mu) - \hat{u}v/(1 + \mu)\big)u,$$

where the operator $A(\mu)$ is defined by

$$A(\mu) = \frac{1}{\varepsilon}\left(\Delta + \xi - \hat{u}\left(1 + 1/(1 + \mu)\right) + \sigma/(\beta + \mu) - \mu\right).$$

The key point now is to show that $A(\mu)$ is invertible for any $\mu \in \Sigma(\mu_0, \theta)$: in fact, this is a sufficient condition for the invertibility of $\mathfrak{A} - \mu$. Now $A(\mu)$ is invertible whenever zero is not an eigenvalue of $A(\mu)$ [6, 7]: to show that this holds true for $\mu \in \Sigma(\mu_0, \theta)$, write the eigenvalue equation for $A(\mu)$

$$A(\mu)\,y(\mu) + \zeta(\mu)\,y(\mu) = 0,$$

$y(\mu)$ being an L^2-normalized eigenfunction. Separating real and imaginary parts of the eigenvalue equation projected on $y(\mu)$, we easily find that

$$\operatorname{Im}\zeta(\mu) \neq 0 \quad \text{whenever} \quad |\operatorname{Im}\mu| > \mu^* = \big((\xi + (\sigma/\beta))/\varepsilon\big)^{1/2}; \tag{3.7}$$

$$\operatorname{Re}\zeta(\mu) > 0 \quad \text{whenever} \quad \operatorname{Re}\mu \geqq 0, \quad \operatorname{Im}\mu \text{ arbitrary}, \tag{3.8}$$

From (3.8) it follows that $\operatorname{Re}\zeta(\mu) > 0$ for $\operatorname{Re}\mu > -\mu_*$, for some $\mu_* > 0$, provided $|\operatorname{Im}\mu|$ is bounded, say, by μ^*. Thus $\zeta(\mu) \neq 0$ for all $\mu \in \mathrm{C}$, excepted possibly the strip $\operatorname{Re}\mu < -\mu^*$, $|\operatorname{Im}\mu| < \mu_*$. It is then clear that $A(\mu)$ is invertible in a sector of type $|\arg(\mu + \mu_0)| < \pi/2 + \theta$, where $0 < \mu_0 < \mu_*$, $0 < \theta < \pi/2$. The second step of the proof is more technical: however, as it requires only minor adaptations of the arguments given in [7], it will be omitted.

Let us now consider the existence and stability of the nontrivial equilibrium solution of (2.5). Both properties can be proved in a very similar way, so that we only list here the most relevant features of the nontrivial equilibrium.

Equilibria are characterized by

$$\Delta u + \xi u - \gamma u^2 - |u|^{\alpha+2} = 0 \quad \text{in } \Omega, \qquad u = 0 \quad \text{in } \partial\Omega, \tag{3.9}$$

and by $v = u$. Now (3.9) has a unique nontrivial solution in $C_0^+(\Omega)$, strictly positive inside Ω. It will be denoted by u.

A remarkable property of $\hat{u}$, which shall be used in the following is, in analogy with (3.5),

$$\langle(\Delta + \xi - \gamma\hat{u} - |\hat{u}|^\alpha\,\hat{u})\varphi, \varphi\rangle \leqq 0, \qquad \varphi_0^1 \in H_0^1(\Omega), \tag{3.10}$$

which implies that the spectrum of $\Delta + \xi - \gamma\hat{u} - |\hat{u}|^\alpha\,\hat{u}$ lies on the non-positive real half-axis.

Remark: Analogous existence and asymptotic stability results can be shown to hold in the spaces $H_0^{p+2} \oplus H^p \oplus H^p$ for (2.3) and $H_0^{p+2} \oplus H^p$ for (2.5), for any $p \geqq 1$. This is true because the maps from $H_0^{p+2} \oplus H^p$ to H_0^p

$$(u, v) \to uv,$$

$$(u, v) \to |u|^\alpha\,uv$$

are continuously FRÉCHET differentiable, as a consequence of the fact that, in our models, the space dimension n is at most 3.

4. Asymptotic expansions

Let us introduce the deviations from the equilibrium: $\varphi = u - \hat{u}$, $\psi = v - \hat{v}$, $\chi = w - \hat{w}$. In terms of these quantities, systems (2.3), (2.5) read

$$\begin{aligned}&\varepsilon\,\partial_t\varphi = (\Delta + \xi - \hat{u})\varphi - \hat{u}\psi + \sigma\chi + R(\varphi, \psi)\\ &\partial_t\psi = \varphi - \psi \qquad\qquad \text{in } (0, \infty) \times \Omega\\ &\partial_t\chi = \varphi - \beta\chi\\ &\varphi = 0 \quad \text{on} \quad (0, \infty) \times \partial\Omega;\\ &\varphi = \Phi = U - \hat{u}, \qquad \psi = \Psi = V - \hat{v}, \qquad \chi = X = W - \hat{w} \quad \text{in } \{0\} \times \Omega.\end{aligned} \tag{4.1}$$

and

$$\begin{aligned} &\varepsilon\, \partial_t\varphi = (\Delta + \xi - \gamma\hat{u} - (\alpha+1)\,|\hat{u}|^\alpha\,\hat{u})\,\varphi - (\gamma\hat{u} + |\hat{u}|^\alpha\,\hat{u})\,\psi + S(\varphi,\psi) \\ &\partial_t\psi = \varphi - \psi \qquad\qquad \text{in } (0,\infty)\times\Omega \\ &\varphi = 0 \quad \text{on } (0,\infty)\times\partial\Omega \\ &\varphi = \Phi = U - \hat{u}\,, \qquad \psi = \Psi = V - \hat{v} \quad \text{in } \{0\}\times\Omega\,. \end{aligned} \tag{4.2}$$

As in the original systems (2.3), (2.5) the nonlinearities define smooth maps from $C_0(\bar{\Omega}) \times C_0(\bar{\Omega})$ to $C_0(\bar{\Omega})$, the terms R, S above are remainders of first-order TAYLOR expansions: in fact,

$$\begin{aligned} &R(\varphi,\psi) = -\varphi\psi\,, \\ &S(\varphi,\psi) = -\gamma\varphi\psi - \left\{(\alpha+1)\int_0^1 |\hat{u} + s\varphi|^\alpha\, ds\right\}(\alpha\varphi^2/2 + \varphi\psi)\,. \end{aligned} \tag{4.3}$$

As we have already observed at the end of Section 3, systems (2.3), (2.5) can be formulated in appropriate direct sums of SOBOLEV spaces. Thus (4.1), (4.2) make sense as well in $H^{p+2} \oplus H^p \oplus H^p$ and $H^{p+2} \oplus H^p$, respectively, and the remainder terms, expressed by formulae (4.3), satisfy

$$|R(\varphi,\psi)|_p\,, \qquad |S(\varphi,\psi)|_p \leqq k_y|\varphi|_{p+2}\,|\psi|_p \tag{4.4}$$

for (φ,ψ) in a neighbourhood Y of $H_0^{p+2} \oplus H^p$, k_y being a positive constant dependent on Y.

For notational convenience, it will be useful to write both systems (4.1), (4.2) synthetically in the form

$$\begin{aligned} &\varepsilon\,\partial_t\varphi = F(\varphi,\zeta) \\ &\partial_t\zeta = I\varphi - C\zeta \end{aligned} \quad \text{in } (0,\infty)\times\Omega\,, \\ \begin{aligned} &\varphi = 0 \quad \text{in } (0,\infty)\times\partial\Omega \\ &\varphi = \Phi\,, \qquad \zeta = Z \quad \text{in } \{0\}\times\Omega\,, \end{aligned} \tag{4.5}$$

where $\zeta = \text{col}\,(\psi,\chi)$ (respectively, $\zeta = \psi$); I is the identity matrix and C a positive diagonal matrix (of the appropriate dimensions), F the semilinear operator defined by the first lines in (4.1), (4.2).

Lemma 4.1.: *The equation $F(\varphi,\zeta) = 0$ is uniquely solvable for φ when (φ,ζ) is in a suitable $H_0^{p+2} \oplus H^p \oplus H^p$ (respectively, $H_0^{p+2} \oplus H^p$) neighbourhood $U \times Z$ of $(0,0)$, for all $p \geqq 0$.*

Proof: A straightforward consequence of the implicit function theorem: in fact, $\hat{F}_\varphi =: F_\varphi(0,0,0)$, i.e.:

$$\Delta + \xi - \hat{u} \quad \text{in case of (4.1)}\,,$$

$$\Delta + \xi - \gamma\hat{u} - (\alpha+1)\,|\hat{u}|^\alpha\,\hat{u} \quad \text{in case of (4.2)}$$

is an isomorphism from H_0^{p+2} to H^p, for all $p \geqq 0$. This is in turn a consequence of formulae (3.5) and (3.10).

We shall now look for power series expansions for the solutions of (4.5) in the following form [2, 3, 5]:

$$\begin{aligned} &\varphi(t,x) = \Sigma\,\varepsilon^k\bar{\varphi}_k(t,x) + \Sigma\,\varepsilon^k\tilde{\varphi}_k(t/\varepsilon,x) \\ &\zeta(t,x) = \Sigma\,\varepsilon^k\bar{\zeta}_k(t,x) + \Sigma\,\varepsilon^k\tilde{\zeta}_k(t/\varepsilon,x)\,. \end{aligned} \tag{4.6}$$

In the above formulae the first expansion — involving the time variable t — is referred to as *outer expansion*, and accounts for the large time behavior of the system. The other expansion, involving the fast time variable t/ε is called *initial* or *transient correction*, as it represents a correction needed to take care of the evolution in an initial time interval (transient), whose length is of the order of ε. The quantities $\bar{\varphi}_k$, $\tilde{\varphi}_k$, $\bar{\zeta}_k$, $\tilde{\zeta}_k$ are determined by solving the differential equations obtained by plugging (4.6) into (4.5), and equating the coefficients of equal ε-powers separately for the t-dependent and for the $\tau = t/\varepsilon$ dependent quantities. These equations should be supplemented by the same boundary conditions an in (4.5); as to the initial conditions, they should be chosen according to the following prescription

$$\begin{aligned} &\bar{\varphi}_k(0,x) + \tilde{\varphi}_k(0,x) = \delta_{k,0}\Phi(x) \\ &\bar{\zeta}_k(0,x) + \tilde{\zeta}_k(0,x) = \delta_{k,0}Z(x) \end{aligned} \quad (\text{for } k = 0, 1, 2, \ldots)\,, \tag{4.8}$$

which is a consequence of the fact that the initial data of (2.3), (2.5) — and therefore of (4.5) — are ε-independent.

In addition, we shall require

$$\begin{aligned} &\lim_{\zeta\to\infty} \varphi_k(\tau,\cdot) = 0 \quad \text{in } H_0^{p+2} \\ &\lim_{\tau\to\infty} \zeta_k(\tau,\cdot) = 0 \quad \text{in } H^p\,. \end{aligned} \tag{4.9}$$

The above conditions, expressing the decay to zero in the fast time scale τ of $\tilde{\varphi}_k$, $\tilde{\zeta}_k$, allows one to interpret these quantities as "transient" corrections.

Let us now write down the equations governing the lowest order of approximation:

$$\begin{aligned} &0 = F(\bar{\varphi}_0,\bar{\zeta}_0) \\ &\partial_t\bar{\zeta}_0 = I\bar{\varphi}_0 - C\zeta_0 \end{aligned} \quad \text{and} \quad \begin{aligned} &\partial_\tau\tilde{\varphi}_0 = \tilde{F}(\tilde{\varphi}_0,\zeta_0) \\ &\partial_\tau\tilde{\zeta}_0 = 0\,. \end{aligned} \tag{4.10}$$

To satisfy the second requirement of (4.9), we see that $\tilde{\zeta}_0$ has to be identically zero — in particular, $\tilde{\zeta}_0(0,\cdot) = 0$, and the equation for $\tilde{\varphi}_0$ has the form

$$\partial_\tau\tilde{\varphi}_0 = F(\tilde{\varphi}_0, 0)\,.$$

We can now prove that the first condition in (4.9) is fulfilled for any initial datum $\tilde{\varphi}_0(0,\cdot)$:

Lemma 4.2: *The equation* (4.11) *together with Dirichlet homogeneous boundary conditions has, for any* $\tilde{\varphi}_0(0,\cdot) \in H_0^p$, *a unique solution which converges to zero in* H_0^{p+2} *as* τ *goes to infinity.*

Proof: In case of (4.1), (4.11) reads

$$\partial_\tau \tilde{\varphi}_0 = (\Delta + \xi - \hat{u})\, \tilde{\varphi}_0 \,. \tag{4.11}$$

Now, the semigroup $G(t)$ generated by $(\Delta + \xi - \hat{u})$ maps, for $t > 0$, H^p into H_0^{p+2} (for any $p \geq 0$); on the other hand, $G(t)$ is an (analytic) semigroup of strictly negative type in H_0^q for all q, whence the conclusion follows. (To see that $G(t)$ is of strictly negative type on H_0^q, observe first that this is obviously true in $H_0 = L^2$, because of (3.5); then notice that $\langle u, (-(\Delta + \xi - \hat{u})^q u\rangle^{1/2}$ is an equivalent norm in H_0^q: application of $(\Delta + \xi - \hat{u})^q$ to (4.11) yields the wanted result). In case of (4.2) the proof is similar.

We can now summarize the procedure by which the appropriate initial data for $\bar{\varphi}_0, \tilde{\varphi}_0, \bar{\zeta}_0, \tilde{\zeta}_0$ should be chosen: start from the initial data (Φ, Z), which are prescribed functions, such that Z is in the neighbourhood $\mathcal{Z}$ of 0 where $F(\varphi, \zeta) = 0$ is solvable for φ. Then define $\tilde{\zeta}_0(0,\cdot) = 0$, and $\bar{\zeta}_0(0,\cdot) = Z$: the second requirements in (4.8) and (4.9) are thus satisfied. Using Lemma 4.1, we can now define $\bar{\varphi}_0(0,\cdot)$ as the unique solution of $F(\bar{\varphi}_0, (0\cdot), \bar{\zeta}_0(0,\cdot)) = 0$, and $\tilde{\varphi}_0(0,\cdot)$ as $\Phi(\cdot) - \bar{\varphi}_0(0,\cdot)$: thus the first requirement of (4.8) is fulfilled, and, by Lemma 4.2, that of (4.9) as well.

To complete our analysis of the lowest order approximation, it remains to investigate the behavior of the outer solution $\bar{\varphi}_0, \bar{\zeta}_0$, which is governed by the first system in $(4.10)_0$. The latter should be regarded as an evolution problem where only the initial datum for $\bar{\zeta}_0$ is prescribed.

In fact, we have:

Theorem 4.3: *The problem*

$$\begin{aligned} &0 = (\Delta + \xi - \hat{u})\,\varphi - \hat{u}\psi + \sigma\chi - \varphi\psi \\ &\partial_t\psi = \varphi - \psi \qquad\qquad \text{in} \quad (0,\infty) \times \Omega \\ &\partial_t\chi = \varphi - \beta\chi \\ &\varphi = 0 \qquad \text{in} \quad (0,\infty) \times \partial\Omega \\ &\psi = \Psi, \qquad \chi = X \text{ in } \quad \{0\} \times \Omega, \end{aligned} \tag{4.12}$$

has, for any (Ψ, X) *in a suitable* $H^p \oplus H^p$ *neighbourhood of* $(0,0)$, *a unique global solution* col (φ, ψ, χ), *such that for all* $t > 0$ $(\varphi(t,\cdot), \chi(t,\cdot))$ *remains in the same neighbourhood and* $\varphi(t,\cdot)$ *remains in a fixed* H_0^{p+2}*-neighbourhood of* 0. *In addition,* $\lim_{t\to\infty} (\psi(t,\cdot), \chi(t,\cdot)) = (0,0)$ *in* $H^p \oplus H^p$ *and* $\lim \varphi(t,\cdot) = 0$ *in* H_0^{p+2}.

A parallel result holds for the problem

$$\begin{aligned} &0 = (\Delta + \xi - \gamma\hat{u} - (\alpha+1)\,|\hat{u}|^\alpha\,\hat{u})\,\varphi - (\gamma\hat{u} + |\hat{u}|^\alpha\,\hat{u})\,\psi + S(\varphi,\psi) \quad \text{in} \quad (0,\infty)\times\Omega \\ &\partial_t\psi = \varphi - \psi \\ &\varphi = 0 \quad \text{in } (0,\infty) \times \partial\Omega; \qquad \psi = \Psi \quad \text{in } \{0\} \times \Omega\,. \end{aligned} \tag{4.13}$$

Proof: Writing again (4.12), (4.13) as the single equation $(4.10)_0$, we first observe that local existence of an unique (smooth) solution holds. In fact, the operator $\mathcal{K}$:

$$(\mathcal{K}\zeta)\,(t) = \exp\,(-Ct)\,\zeta(0) + \int_0^t \exp\,(-C(t-s))\,\varphi(\zeta(s))\,\mathrm{d}s$$

where $\varphi(\zeta(s))$ is the unique solution of $F(\varphi, \zeta(s)) = 0$ for $\zeta(s) \in \mathcal{Z}$, maps, for T sufficiently small, $C([0,T]; \mathcal{Z})$ into itself, and it is a contraction.

To proceed further, observe that we can write $\varphi(\zeta(s))$ as $-\hat{F}_\varphi^{-1}\hat{F}_\zeta\zeta + Q(\zeta)$ (where "^" means that derivatives are evaluated at $\varphi = \zeta = 0$), and Q verifies $|Q(\zeta)|_{p+2} = o(|\zeta|_p)$. Thus the full evolution is described by a single equation for ζ, i.e.:

$$\partial_t\zeta = -\,I\hat{F}_\varphi^{-1}\hat{F}_\zeta\zeta + Q(\zeta) \tag{4.14}$$

where $|Q(\zeta)|_{p+2} = o(|\zeta|_p)$.

A linearization argument will then provide the desired result, once we know that the linear part

$$-I\hat{F}_\varphi^{-1}\hat{F}_\zeta$$

generates a strongly continuous semigroup on $H^p \oplus H^p$, (respectively, on H^p), of strictly negative type. This property of the linear part enables in fact to show [4, 7] that $(0,0)$ is asymptotically stable for (4.14), whence, by the implicit function theorem, the corresponding property for the first component $\varphi(t,\cdot)$ easily follows.

It remains therefore to show that the linear part of (4.14) actually generates a semigroup of strictly negative type. To do this, we shall investigate the asymptotic behavior of the solutions of the corresponding (linear) system: specifically, we shall first consider the case of (4.12), which leads to the system

$$\begin{aligned} &0 = (\Delta + \xi - \hat{u})\,\varphi - \hat{u}\psi + \sigma\chi \\ &\partial_t\psi = \varphi - \psi \\ &\partial_t\chi = \varphi - \beta\chi\,. \end{aligned} \tag{4.15}$$

The smoothness of the (local) solution of this system enables one to take time derivative of the first equation and to interchange time derivatives with Laplacians: expressing then ψ_t and χ_t by means of the second and third equation, we get

$$0 = (\Delta + \xi - \hat{u})\,\varphi_t - \hat{u}\varphi + \hat{u}\psi + \sigma\varphi - \sigma\beta\chi\,.$$

Writing $\hat{u}$ as $\beta\hat{u} + (1-\beta)\hat{u}$, and recalling the first equation in (4.15), we get

$$0 = A\varphi_t + B\varphi - (1-\beta)\,\hat{u}\psi\,, \tag{4.16}$$

where $A = -(\Delta + \xi - \hat{u})$, $B = -\beta(\Delta + \xi + (\sigma/\beta) - (1 + 1/\beta)\,\hat{u})$ are strictly positive selfadjoint operators. Using now the second equation in (4.15), and multiplying scalarly (4.16) by φ we obtain

$$0 = (1/2)\,\langle A\varphi, \varphi\rangle_t + \langle B\varphi, \varphi\rangle + (\beta - 1)\,(1/2)\,[\langle\hat{u}, \psi^2\rangle_t + \langle\hat{u}, \psi^2\rangle]\,.$$

As the operators of the second order A and B satisfy, for a suitable constant $c > 0$:

$$\langle Bu, u\rangle \geqq c\langle Au, u\rangle\,, \qquad u \in H_0^1\,,$$

we eventually get

$$(\mathrm{d}/\mathrm{d}t)\,\{(1/2)\,(\langle A\varphi, \varphi\rangle + (\beta-1)\,\langle\hat{u}, \psi^2\rangle)\} \leqq -c\langle A\varphi, \varphi\rangle - (\beta - 1)\,\langle\hat{u}, \psi^2\rangle\,.$$

As, by assumption, $\beta \geqq 1$, we may conclude that, as Ψ, $X \in H^1$, the quantity $\langle A\varphi, \varphi\rangle$ converges to zero exponentially as $t \to \infty$ (and the same holds, if $\beta > 1$, for $\langle\hat{u}, \psi^2\rangle$): in other words, φ converges in H_0^1 to zero: using the second and third equations in (4.15), the convergence to zero of ψ, χ in H^1 follows, whence, using again the first equation, φ turns out to converge to zero in H_0^3. So far, the proof is complete if $p = 1$: if $p > 1$, we proceed in the same way, using first the last two equations, then the first one, and possibly repeating this procedure a finite number of times.

A similar method applies to system (4.13): alternatively, the linearized system can be written as

$$A\psi_t = -(\gamma\hat{u} + |\hat{u}|^\alpha\,\hat{u})\,\psi - A\psi$$

where $A = -(\Delta + \xi - \gamma\hat{u} - (\alpha + 1)\,|\hat{u}|^\alpha\,\hat{u})$, so that

$$(1/2)\,(\mathrm{d}/\mathrm{d}t)\,\langle A\psi, \psi\rangle \leqq -\langle A\psi, \psi\rangle\,,$$

thus proving the convergence to zero of ψ in H^1. Using the first equation, φ is seen to converge in H_0^3 to zero, and so on.

Concluding, we have seen that the linear part of (4.14) gives rise to an exponentially decaying semigroup in $H^p \oplus H^p$ (in H^p, respectively): thus $\zeta = 0$ is asymptotically stable for the full system (4.14), and the proof is complete.

References

1 Canosa, J., Diffusion in nonlinear multiplicative media, J. Math. Phys. **10** (1969), 1862 –1868.
2 Eckhaus, W., Matched Asymptotic Expansions and Singular Perturbations, American Elsevier, New York 1973.
3 Hoppensteadt, F., Properties of solutions of ordinary differential equations with small parameters, Commun. Pure Appl. Math **24** (1971), 807–840.
4 Iooss, G., Topics in Bifurcation of Maps and Applications, Minneapolis, University of Minnesota (Lecture Notes) 1977.
5 Mika, J., Singularly perturbed evolution equations in Banach spaces, J. Math. Analysis Appl. **58** (1977), 189–201.
6 de Mottoni, P.; Talenti, G.; Tesei, A., Asymptotic stability results for a class of non-linear parabolic equations, Ann. di Mat. Pura Appl., IV, Ser. **115** (1977), 295–310.
7 de Mottoni, P.; Tesei, A., Asymptotic stability results for a system of quasilinear parabolic equations, Applicable Analysis **9** (1979), 7–21.
8 de Mottoni, P.; Tesei, A., A Singular Perturbation Approach for a Rection Diffusion System arising in Nuclear Reactor Theory, Mathem. Methods in the Appl. Sciences, **2** (1980), 91–107.
9 Reister, D. D., Bounds for the Solutions of Nonlinear ParabolicEquations, Ph. D. Thesis, Nucl. Engin. Depart., University of California at Berkeley 1969.
10 Reister, D., Optimum bounds for space-time reactor problems using linear programming, Nucl. Sci. and Engineering **46** (1971), 197–202.
12 Rumble III, E. T.; Kastenberg, W. E., On the application of eigenfunction expansion to nonlinear space-time reactor dynamics, Nucl. Sci. and Engineering **49** (1972), 172–187.
13 Shampine, L. F.; Gear, C. W.; A user's view of solving stiff ordinary differential equations, Siam Review **21** (1979), 1–17.
14 Tikhonov, A. N., Systems of differential equations containing a small parameter, Mat. Sbornik **31** (1952), 575–586.
15 Vasil'eva, A. V., Nonlinear differential equations containing a small parameter, Russ. Math. Surv. **18** (1963), 13–84.

Eingereicht am 20. 2. 1980

Anschrift: Professor Dr. P. de Mottoni; Professor Dr. A. Tesei, Consiglio Nazionale delle Richerche, Istituto per le Applicazioni del Calcolo (IAC) "Mauro Picone", Viale del Policlinico, 137, I-00161 Roma, Italien

ZAMM 60, 623 – 629 (1980)

A. SANDQVIST / K. M. ANDERSEN

Some Results Concerning Periodic Solutions of Plane Autonomous Differential Systems

Betrachtet wird das ebene autonome System $\dot{x} = P(x, y)$, $\dot{y} = Q(x, y)$. Damit verbunden ist eine Differentialgleichung in Polarkoordinaten, die untersucht wird, um Aussagen über die mögliche Anzahl von geschlossenen Trajektorien zu gewinnen. Für diese Differentialgleichung werden einige Kriterien dafür hergeleitet, daß sie höchstens eine bzw. zwei 2π-periodische Lösungen hat, womit entsprechende Resultate für die Anzahl der geschlossenen Trajektorien des autonomen Systems vorliegen. Zur Illustration werden die Ergebnisse angewendet, um einen Fall von Bifurkation in einem speziellen Beispiel zu untersuchen.

In order to obtain results concerning the possible number of closed trajectories to the plane autonomous system $\dot{x} = P(x, y)$, $\dot{y} = Q(x, y)$ a related polar differential equation is considered. Some criteria are derived securing this equation to have at most one, respectively at most two 2π-periodic solutions — thereby giving corresponding results for the number of closed trajectories to the autonomous system. As an illustration the results are applied to investigate a case of bifurcation in a specific example.

Рассматривается автономная система $\dot{x} = P(x, y)$, $\dot{y} = Q(x, y)$ в плоскости x, y. Исследуется связанное с данной системой дифференциальное уравнение в полярных координатах, чтобы получить результаты о возможном числе замкнутых траекторий. Получены некоторые критерия, гарантирующие существование у указанного дифференциального уравнения не больше одного либо небольше двух 2π-периодичных решений что одновременно означает верность соответствующих результатов о числе замкнутых траекторий автономной системы. В качестве иллюстрации результаты применены к исследованию одного случая бифуркации в специальном примере.

1. Introduction

Investigating *two-dimensional autonomous systems*

$$\dot{x} = P(x, y) , \qquad \dot{y} = Q(x, y) , \qquad (x, y) \in D , \tag{1}$$

where D is some open domain, it is often useful to introduce polar coordinates. Setting $x = r \cos \theta$, $y = r \sin \theta$ the system (1) is transformed into the equation

$$\frac{dr}{d\theta} = rf(\theta, r) \tag{2}$$

where

$$f(\theta, r) = \frac{\cos \theta P(r \cos \theta, r \sin \theta) + \sin \theta\, Q(r \cos \theta, r \sin \theta)}{-\sin \theta\, P(r \cos \theta, r \sin \theta) + \cos \theta\, Q(r \cos \theta, r \sin \theta)} \tag{3}$$

is 2π-periodic in θ. The solutions $r(\theta)$ of the equation (2) correspond to trajectories $r = r(\theta)$ of the system (1) lying in some component of the set

$$\hat{D} = \{(x, y) \in D \mid r^2\dot{\theta} = -y\, P(x, y) + x\, Q(x, y) \neq 0\} .$$

In particular 2π-*periodic* solutions $r(\theta)$ correspond with *closed* trajectories in $\hat{D}$.

A set A in the (x, y)-plane will be called *annular* if $(0, 0) \notin A$ and if any halfline $(x, y) = (\alpha t, \beta t)$, $t > 0$, intersects A in a nonempty convex set. In the following we only consider the case, where D is an annular open domain, in which $\dot{\theta} \neq 0$ holds (i.e. $\hat{D} = D$). Then the function f will be defined in the open domain

$$\Omega = \{(\theta, r) \mid (r \cos \theta, r \sin \theta) \in D\} ,$$

lying in the halfplane $r > 0$ and having the properties

(i) $(\theta, r_1) \in \Omega \wedge (\theta, r_2) \in \Omega \Rightarrow \forall\, \alpha \in [0, 1]\colon (\theta, (1 - \alpha)\, r_1 + \alpha r_2) \in \Omega$,

(ii) $(\theta, r) \in \Omega \Rightarrow \forall\, p \in \mathbb{Z}\colon (\theta + 2\pi p, r) \in \Omega$.

The purpose of this paper is to derive sufficient conditions that the system (1) has at most one or at most two closed trajectories, i.e. that the equation (2) has at most one or at most two 2π-periodic solutions.

However our considerations apply not only to equations of the special form (2), but as well to equations of the general form

$$\frac{dr}{d\theta} = F(\theta, r) , \tag{4}$$

where F is 2π-periodic in θ. In section 2 we in particular derive some conditions — under weak assumptions — securing uniqueness of 2π-periodic solutions to the equation (4). In section 3 we prove a theorem, stating that under certain conditions an equation (4) has *at most two* 2π-periodic solutions (Theorem 2). It is a common feature of these results that there is involved an auxiliary function $V(\theta, r)$, which in each concrete case has to be chosen in a suitable way — quite analogous to the situation in the LYAPUNOV stability theory. As to the equation (2) — and hence the system (1) — the choice $V(\theta, r) = \log r$ gives the most "natural" results (stated at the end of section 2 and section 3). In the concluding section 4 we give an example, showing how the results can be used for a case of bifurcation in a system (1).

If the functions P and Q are C^1 the well-known classical theorems on uniqueness of closed trajectories to a system (1) will hold. We first mention the criterions of Dulac and Bendixson. The Dulac criterion reads: *if there exists a C^1-function $B(x, y)$, $(x, y) \in D$, such that the divergence* div (BP, BQ) *in the open annular domain D is of fixed sign [except for a subset of empty interior], then the system* (1) *cannot have more than one closed trajectory in D.* For a proof, see e.g. [1] p. 333. Setting $B = 1$, we get the Bendixson criterion as a special case. Next we mention the criterion of the characteristic exponent. Recall that the characteristic exponent μ of a given closed trajectory Γ: $(x, y) = (\varphi(t), \psi(t))$, $t \in \mathbb{R}$, of period T is given by

$$\mu = \frac{1}{T}\int_0^T \operatorname{div}(P, Q) \circ (\varphi(t), \psi(t))\, dt .$$

If μ is different from zero, then Γ is isolated and orbitally asymptotically stable, positively if $\mu < 0$ and negatively if $\mu > 0$ [see e.g. [4] p. 222f]. Moreover two adjacent closed trajectories cannot both be positively — or negatively — orbitally asymptotically stable on the sides facing one another [see e.g. [3] p. 397]. From these propositions the criterion follows: *if the characteristic exponent of any closed trajectory in the open annular domain D must be positive, then the system* (1) *cannot have more than one closed trajectory in D.* In this statement the word *positive* can be substituted by *negative.*

The last criterion corresponds to a well-known similar criterion for equations (2). If namely the characteristic exponent of the solution $r(\theta)$ to (2) corresponding to Γ above is denoted by σ, it can be shown by direct calculation that σ and μ are connected by the formula (see Appendix A)

$$2\pi\sigma = \pm T\mu , \tag{5}$$

where the sign $+$ or $-$ on the right hand side is to be chosen in accordance with the sign of $\dot\theta$ in D. Recall that σ is given by

$$\sigma = \frac{1}{2\pi}\int_0^{2\pi} \frac{\partial}{\partial r}\,(rf(\theta, r))|_{r=r(\theta)}\, d\theta . \tag{6}$$

The relation (5) allows us to derive another condition for uniqueness. From (6) it follows that

$$\sigma = \frac{1}{2\pi}\int_0^{2\pi} f(\theta, r(\theta))\, d\theta + \frac{1}{2\pi}\int_0^{2\pi} r(\theta)\, f_r'(\theta, r(\theta))\, d\theta = \frac{1}{2\pi}\int_0^{2\pi} r(\theta)\, f_r'(\theta, r(\theta))\, d\theta$$

since $f(\theta, r(\theta)) = (d/d\theta) \log r(\theta)$. We conclude that $\sigma \geqq 0$ holds throughhout if f_r' is nonnegative. Furthermore $\sigma > 0$ must hold if the set $\{(\theta, r) \mid f_r'(\theta, r) > 0\}$ contains some point of any curve $r = \varphi(\theta)$, $\theta \in \mathbb{R}$ (since $\sigma = 0$ implies $f_r'(\theta, r(\theta)) = 0$ for $\theta \in [0, 2\pi]$, i.e. for $\theta \in \mathbb{R}$). Under these circumstances a 2π-periodic solution of the equation (2) — and correspondingly by (5) a closed trajectory of the system (1) — must be unique. Of course an analoguous result holds, if the inequality signs as to f_r' are reversed. In section 2 it will be shown that the conditions involving f_r' in this criterion may be replaced by weaker conditions.

Now the above propositions only apply to problems in which the involved functions P, Q and hence f are C^1, whereas the results of section 2 for an equation (4) only need F to be continuous.

2. Some results on uniqueness

Let Ω be an open domain in the Euclidian (θ, r)-plane with the properties (i) and (ii) of section 1. We consider the equation

$$\frac{dr}{d\theta} = F(\theta, r) , \qquad (\theta, r) \in \Omega , \tag{7}$$

where F is continuous in Ω and 2π-periodic in θ.

If V is any C^1-function in Ω, the derivative of V with respect to (7) is the continuous function defined by

$$V^*(\theta, r) = V_r'(\theta, r)\, F(\theta, r) + V_\theta'(\theta, r) .$$

From this it follows that

$$\frac{d}{d\theta} V(\theta, r(\theta)) = V^*(\theta, r(\theta))$$

holds for any solution $r(\theta)$ to (7). In particular — if V is 2π-periodic in θ — the equation

$$\int_0^{2\pi} V^*(\theta, r(\theta))\, d\theta = 0 \tag{8}$$

holds for any 2π-periodic solution $r(\theta)$.

Suppose that $r = r_1(\theta)$ and $r = r_2(\theta)$ are two different 2π-periodic solutions to (7). Then

$$\hat{r}_1(\theta) = \min\{r_1(\theta), r_2(\theta)\}, \qquad \hat{r}_2(\theta) = \max\{r_1(\theta), r_2(\theta)\} \tag{9}$$

define two 2π-periodic solutions to (7) with $\hat{r}_1(\theta) \leq \hat{r}_2(\theta)$ for all θ, and

$$\hat{\Omega} = \{(\theta, r) \mid \hat{r}_1(\theta) \leqq r \leqq \hat{r}_2(\theta), \theta \in \mathbb{R}\} \quad . \tag{10}$$

is a closed and connected subset of Ω — not necessarily a domain. From (8) we easily get the following result:

Theorem 1: *Suppose there exists a C^1-function V in Ω, 2π-periodic in θ, such that the derivative $V^*(\theta, r)$ with respect to* (7) *is weakly increasing (decreasing) in r for any fixed θ. If $r_1(\theta)$ and $r_2(\theta)$ are two different 2π-periodic solutions and $\hat{\Omega}$ is given by* (9) *and* (10), *then*

$$V^*(\theta, r) = V^*(\theta, s) \quad \textit{for all} \quad (\theta, r), (\theta, s) \in \hat{\Omega},$$

and any solution $r(\theta)$, $\theta \in \mathbb{R}$, to (7) *entirely lying in $\hat{\Omega}$ satisfies the relation*

$$V(\theta, r(\theta)) = c + \int_0^\theta V^*(\tau, r_1(\tau))\, d\tau, \theta \in \mathbb{R}, \tag{11}$$

for some constant c.

From the first statement of the theorem we immediately get:

Corollary 1: *If in particular V^* is strictly monotonic in r for all fixed θ, there will be at most one 2π-periodic solution to the equation* (7).

Another immediate consequence is the following slight extension of the Théorème XIX in the POINCARÉ Œuvres [5].

Corollary 2: *If in particular $F(\theta, r)$ satisfies conditions assuring local uniqueness of solutions to* (7) *and besides $V^*(\theta, r)$ is strictly monotonic in r for at least one $\theta \in \mathbb{R}$, the equation* (7) *has at most one 2π-periodic solution.*

In the next corollary there is also involved a condition concerning the function V itself.

Corollary 3: *If in particular $V'_r(\theta, r) \neq 0$ holds in Ω for almost all θ* (i.e. *the θ-coordinates for the zeros of $V'_r(\theta, r)$ form a set of measure zero*) *then any solution $r(\theta)$, $\theta \in \mathbb{R}$, to* (7) — *entirely lying in $\hat{\Omega}$ — will be 2π-periodic and the expression* (11) *determines the complete solution to* (7) *in $\hat{\Omega}$. Moreover, $\hat{\Omega}$ will be a closed domain in the interior of which there is local uniqueness of solutions to* (7).

Proof: Suppose now that V'_r satisfies the stated condition. This means that $V(\theta, r)$ is one to one in r for almost all θ. From (8) it follows that the right hand side of (11) is 2π-periodic — hence $V(\theta, r(\theta + 2\pi)) = V(\theta, r(\theta))$ for all θ and then $r(\theta + 2\pi) = r(\theta)$ for almost all θ. By continuity the periodicity of $r(\theta)$ follows. If a C^1-function $s(\theta)$, $\theta \in \mathrm{R}$ — entirely lying in $\hat{\Omega}$ — for some constant c satisfies the relation

$$V(\theta, s(\theta)) = c + \int_0^\theta V^*(\tau, r_1(\tau))\, d\tau, \qquad \theta \in \mathbb{R},$$

we get by differentiation

$$V'_r(\theta, s(\theta)) \cdot s'(\theta) = V^*(\theta, r_1(\theta)) - V'_\theta(\theta, s(\theta)) = V^*(\theta, s(\theta)) - V'_\theta(\theta, s(\theta)) = V'_r(\theta, s(\theta))\, F(s(\theta), \theta),$$

from which $s'(\theta) = F(s(\theta), \theta)$ follows for almost all θ, i.e. $s(\theta)$ must be a solution to (7). Moreover the solutions correspond to the constants. If namely $s_1(\theta)$ and $s_2(\theta)$ are two solutions of (7) in $\hat{\Omega}$ satisfying (11) with the same constant c, we get $V(\theta, s_1(\theta)) = V(\theta, s_2(\theta))$ for all $\theta \in \mathbb{R}$. By the same type of argument as above we conclude that $s_1 = s_2$.

Suppose that $r_1(\theta_0) = r_2(\theta_0)$ holds for some $\theta_0 \in \mathbb{R}$. Then any solution $s(\theta)$ in $\hat{\Omega}$ will correspond to the same constant, namely

$$c = V(\theta_0, r_1(\theta_0)) - \int_0^{\theta_0} V^*(\tau, r_1(\tau))\, d\tau .$$

Hence $r_1(\theta) = r_2(\theta)$ must hold for all θ, contradicting that these two solutions are assumed to be different. So we have $r_1(\theta) \neq r_2(\theta)$ for all $\theta \in \mathbb{R}$, and $\hat{\Omega}$ is a closed domain. The last arguments show that there is local uniqueness of solutions in the interior of $\hat{\Omega}$.

In the case of local uniqueness of solutions to (7) the set $\hat{\Omega}$ will (a priori) be a closed domain and from the first statement of theorem 1 it follows that $(V^*)'_r$ must exist and be zero in the open "band" consisting of the interior of $\hat{\Omega}$. Hence any further condition excluding this property of V^* yields a uniqueness theorem.

As an example we return to the equation (2), i.e. $F(\theta, r) = rf(0, r)$ where f is a continuous function and 2π-periodic in θ, and where r is positive. Choosing $V(r, \theta) = \log r$, we get $V^*(\theta, r) = f(\theta, r)$. If $f(r, \theta)$ is weakly increasing (decreasing) in r for any fixed θ, then the complete solution of this equation in the closed set between two given, different 2π-periodic solutions $r_1(\theta)$ and $r_2(\theta)$ will be

$$r(\theta) = K e^{\int_0^\theta f(\tau, r_1(\tau))\, d\tau}, \qquad K \in \mathbb{R}_+ ,$$

according to corollary 3. If in particular $f(\theta, r)$ is strictly monotonic in r for all $\theta \in \mathbb{R}$ a 2π-periodic solution of the equation (2) will be unique (corollary 1). The same will be true, if there is local uniqueness of solutions to the equation and the strict monotonicity only holds for at least one $\theta \in \mathbb{R}$ (corollary 2). This means — as to the corresponding autonomous system (1) — if the function f satisfies the conditions above, then the annular set between two different closed trajectories is the union of closed trajectories homothetic to each other — and in the strictly monotonic case a closed trajectory will be unique.

Note, that we (in view of the remarks after corollary 3) have shown the criterion derived in section 1, but under weaker conditions.

3. A condition for at most two periodic solutions

Our next result needs the functions involved (i.e. F and V) to satisfy further conditions as to differentiability.

Theorem 2: *Suppose that F is a C^1-function. The equation (7) has at most two 2π-periodic solutions if there exists a C^2-function V in Ω, 2π-periodic in θ, such that*

(a) *$V_r'(\theta, r) \neq 0$ holds in Ω;*

(b) *$1/V_r'(\theta, r) \cdot (V^*)_r'(\theta, r)$ is weakly increasing (decreasing) for any fixed θ — and strictly for at least one θ.*

Proof: If $r_1(\theta)$, $r_2(\theta)$ are two different periodic solutions — say $r_1(\theta) < r_2(\theta)$ — then the function

$$g(\theta) = V(\theta, r_1(\theta)) - V(\theta, r_2(\theta))$$

is nonzero for any θ because of (a), and it is 2π-periodic. Moreover

$$g'(\theta) = V^*(\theta, r_1(\theta)) - V^*(\theta, r_2(\theta)) .$$

From the second mean value theorem it follows that there exists a $\varrho(\theta) \in]r_1(\theta), r_2(\theta)[$ such that

$$\frac{g'(\theta)}{g(\theta)} = \frac{(V^*)_r'(\theta, \varrho(\theta))}{V_r'(\theta, \varrho(\theta))} .$$

Hence

$$\int_0^{2\pi} \frac{(V^*)_r'(\theta, \varrho(\theta))}{V_r'(\theta, \varrho(\theta))} \, d\theta = 0 .$$

If there exists a third periodic solution $r_3(\theta)$ $\big(\text{let } r_2(\theta) < r_3(\theta)\big)$ we get

$$\int_0^{2\pi} \frac{(V^*)_r'(\theta, \tau(\theta))}{V_r'(\theta, \tau(\theta))} \, d\theta = 0$$

for some $\tau(\theta) \in]r_2(\theta), r_3(\theta)[$, i.e. $\varrho(\theta) < \tau(\theta)$ for all θ. This contradicts the assumption (b).

Let again $F(\theta, r) = rf(\theta, r)$, where now f is a C^1-function in Ω, and choose again $V(\theta, r) = \log r$. Then $V_r'(\theta, r) = 1/r > 0$ and $V^*(\theta, r) = f(\theta, r)$. Hence the statement of Theorem 2 holds if $rf_r'(\theta, r)$ is weakly increasing (decreasing) for any fixed θ — strictly for at least one θ. Suppose now that this condition is fulfilled and that the equation (7) has two different 2π-periodic solutions $r_1(\theta)$, $r_2(\theta)$, say $r_1(\theta) < r_2(\theta)$. From the proof above it follows that

$$\int_0^{2\pi} \varrho(\theta) f_r'(\theta, \varrho(\theta)) \, d\theta = 0$$

for some $\varrho(\theta) \in]r_1(\theta), r_2(\theta)[$. If σ_1 and σ_2 denote the characteristic exponents respectively for the two solutions we get from (6)

$$\sigma_1 < \frac{1}{2\pi} \int_0^{2\pi} \varrho(\theta) f_r'(\theta, \varrho(\theta)) \, d\theta < \sigma_2$$

if $rf_r'(\theta, r)$ is increasing, and the reversed inequalities if $rf_r'(\theta, r)$ is decreasing. We conclude that σ_1 and σ_2 have opposite signs. We have shown

Theorem 3: *Suppose that f is a C^1-function in $\{(r, \theta) \in \Omega \mid r > 0\}$, 2π-periodic in θ, such that*

$$rf_r'(\theta, r) \text{ is weakly increasing (decreasing) for any fixed } \theta \text{ — and strictly for at least one } \theta . \tag{12}$$

Then the equation (7) with $F(\theta, r) = rf(\theta, r)$ has at most two 2π-periodic solutions. If there are in fact two such solutions then they will both be asymptotically stable, one for $\theta \to +\infty$, the other for $\theta \to -\infty$.

Note that the condition on $rf_r'(\theta, r)$ at first glance leads one to believe that there might be three 2π-periodic solutions with $\sigma < 0$, $\sigma = 0$ and $\sigma > 0$, respectively.

Remarks: Returning to the system (1) we get a result about closed trajectories if P and Q are C^1-functions in some open annular domain D surrounding — but not including — the origin in the (x, y)-plane, and if

$$R(x, y) = -yP(x, y) + xQ(x, y) \neq 0 \quad \text{in } D .$$

By direct calculation it can be shown (see Appendix B) that

$$rf'_r(\theta, r) = (x^2 + y^2) \operatorname{div} \frac{1}{R} (P, Q) . \tag{13}$$

If the condition (12) of Theorem 3 is satisfied, the system (1) has at most two closed trajectories in D. If there are in fact two, we may conclude — in view of (5) — that they will both be asymptotically orbitally stable from both sides, the one positively — the other negatively.

Another result, derived from the above one, is the following. *Let K be some closed, annular* LEVINSON-SMITH *cell — then there exists in K at least one closed trajectory to the system* (1) (see e.g. [2] p. 165f). *If the function*

$$\frac{\partial}{\partial r}[rf'_r(\theta, r)] = \frac{\partial}{\partial r} r^2 \operatorname{div} \frac{1}{R(x, y)} \left(P(x, y), Q(x, y)\right) \Big|_{\substack{x = r\cos\theta \\ y = r\sin\theta}}$$

has fixed sign in some open set containing K, then there will be exactly one closed trajectory in K.

4. An example

To demonstrate the use of theorem 3 for two-dimensional systems we consider the case

$$\dot{x} = P(x, y; \varepsilon) = y - x[(x^2 + y^2 - \varrho^2)^2 - \varepsilon H(x, y)] ,$$
$$\dot{y} = Q(x, y; \varepsilon) = -x - y[(x^2 + y^2 - \varrho^2)^2 - \varepsilon H(x, y)] ,$$

where ϱ is a positive constant, ε a parameter and $H(x, y)$ any C^2-function in the entire (x, y)-plane — only assumed to be positive on the circle $x^2 + y^2 = \varrho^2$. We have $R(x, y) = -(x^2 + y^2) < 0$ in the open annular domain $D = \mathbb{R}^2 \setminus \{(0, 0)\}$. The trajectories in this domain are given by

$$\frac{dr}{d\theta} = rf(\theta, r; \varepsilon) ,$$

where $f(\theta, r; \varepsilon) = (r^2 - \varrho^2)^2 - \varepsilon h(\theta, r)$. We have set $h(\theta, r) = H(r\cos\theta, r\sin\theta)$.

If $\varepsilon = 0$, the system has exactly one closed trajectory, namely the circle $r = \varrho$, which is orbitally semistable. We investigate the behaviour as to closed trajectories when ε varies around zero.

If $\varepsilon < 0$, the function $f(\theta, r, \varepsilon)$ will be positive in some annular domain $\varrho_1 < r < \varrho_2$ around the circle $r = \varrho$, since $h(\theta, \varrho) > 0$ for all θ by assumption. Clearly the annular domain contains no closed trajectory.

If $\varepsilon > 0$, we have $\dot{r} = \varrho\varepsilon h(\theta, \varrho) < 0$ on the circle $r = \varrho$. Moreover $\dot{r} = -r[(r^2 - \varrho^2)^2 - \varepsilon h(\theta, r)] < 0$ holds on any circle $r = R$ $(R \neq \varrho)$ provided ε is small enough — say $0 < \varepsilon < \varepsilon_R$. Hence the closed, annular domain K_R between the two circles $r = \varrho$ and $r = R$ $(R \neq \varrho)$ is a LEVINSON-SMITH cell containing at least one closed trajectory — if $0 < \varepsilon < \varepsilon_R$. By calculation we get

$$f'_r(\theta, r; \varepsilon) = 4r(r^2 - \varrho^2) - \varepsilon \frac{\partial}{\partial r} h(\theta, r) .$$

The sign of this function in any neighbourhood of the circle $r = \varrho$ is completely unknown (without further conditions on the derivative $\partial h/\partial r$). Hence we cannot use the results at the end of section 2 to conclude uniqueness. However, theorem 3 will allow us to do so. By calculation we get

$$\frac{\partial}{\partial r}[rf'_r(\theta, r, \varepsilon)] = 16r\left(r^2 - \frac{\varrho^2}{2}\right) - \varepsilon \frac{\partial}{\partial r}\left[r \frac{\partial}{\partial r} h(\theta, r)\right].$$

If $R > \frac{1}{\sqrt{2}} \varrho$ this function will be positive in the domain K_R if ε is small enough — say $\varepsilon < \hat{\varepsilon}_R$. From the remarks at the end of section 3 it follows that K_R contains exactly one closed trajectory-provided $R > \frac{1}{\sqrt{2}} \varrho$ and $0 < \varepsilon < \min \{\varepsilon_R, \hat{\varepsilon}_R\}$. Since R may be taken in each of the two intervals $\left]\frac{1}{\sqrt{2}} \varrho, \varrho\right[$ and $]\varrho, +\infty[$, it follows that this is a case of *bifurcation*: A suddenly arising semistable closed trajectory splits into two closed trajectories, the one positively and the other negatively asymptotically stable, when ε passes the value zero from below.

References

1 ANDRONOV, A. A.; VITT, A. A.; KHAIKIN, S. E., Theory of Oscillators, Pergamon Press, Oxford 1966.
2 CESARI, L., Asymptotic Behavior and Stability Problems in Ordinary Differential Equations, Third Edition, Springer, Berlin-Heidelberg-New York 1971.
3 CODDINGTON, E. A.; LEVINSION, N., Theory of Ordinary Differential Equations, McGraw Hill, New York 1955.
4 HALE, J. K.: Ordinary Differential Equations, Wiley-Interscience, New York 1969.
5 POINCARÉ, H., Oeuvres. Tome I, Gauthiers Villars, Paris 1928.

Eingereicht am 17. 8. 1979, korrigierte Fassung am 21. 3. 1980

Anschrift: M. Sc. A. SANDQVIST, M. Sc. K. M. ANDERSEN, Mathematical Institute, The Technical University of Denmark, Building 303, DK-2800 Lyngby, Denmark.

Appendix A

If the closed trajectory is given by $(x, y) = (\varphi(t), \psi(t))$, $t \in \mathbb{R}$, we have

$$\mu = \frac{1}{T}\int_0^T \operatorname{div}(P, Q) \circ (\varphi(t), \psi(t))\, dt\,,$$

and from section 1 we have

$$\sigma = \frac{1}{2\pi}\int_0^{2\pi} r(\theta) f_r'(\theta, r(\theta))\, d\theta\,.$$

The formula (5) follows from the fact that

$$\int_0^{2\pi} r(\theta) f_r'(\theta, r(\theta))\, d\theta = \pm \int_0^T \operatorname{div}(P, Q) \circ (\varphi(t), \psi(t))\, dt\,, \tag{*}$$

($+$ if $\dot\theta > 0$, $-$ if $\dot\theta < 0$) which will now be shown. To this end we need the formula – proved in Appendix B –

$$f_r'(\theta, r) = r \operatorname{div} \frac{1}{R}(P, Q)\Big|_{\substack{x = r\cos\theta\\ y = r\sin\theta}}$$

where $R(x, y) = -yP(x, y) + xQ(x, y)$. Since

$$\frac{d\theta}{dt} = \frac{1}{r^2} R(x, y)\,, \qquad (x, y) \in \hat{D}\,,$$

is different from zero on the closed trajectory, we may change the variable in the left hand side of (*), getting

$$\begin{aligned}\int_0^{2\pi} r(\theta) f_r'(\theta, r(\theta))\, d\theta &= \pm \int_0^T (x^2 + y^2) \operatorname{div} \frac{1}{R(x, y)} (P(x, y), Q(x, y))\Big|_{\substack{x=\varphi(t)\\ y=\psi(t)}} \frac{d\theta}{dt}\, dt\\ &= \pm \int_0^T R(x, y) \operatorname{div} \frac{1}{R(x, y)} (P(x, y), Q(x, y))\Big|_{\substack{x=\varphi(t)\\ y=\psi(t)}} dt\\ &= \pm \int_0^T \operatorname{div}(P, Q) \circ (\varphi(t), \psi(t))\, dt \mp \int_0^T \frac{1}{R(x, y)} (P(x, y), Q(x, y)) \circ \operatorname{grad} R(x, y)\Big|_{\substack{x=\varphi(t)\\ y=\psi(t)}} dt\,,\end{aligned}$$

since $\operatorname{div} \frac{1}{R}(P, Q) = \frac{1}{R} \operatorname{div}(P, Q) - \frac{1}{R^2}(P, Q) \cdot \operatorname{grad} R$.

So we only have to show that the second member of the difference above is zero. Using

$$(P, Q) \cdot \operatorname{grad} R = \varphi'(t) \frac{\partial R}{\partial x} + \psi'(t) \frac{\partial R}{\partial y} = \frac{d}{dt} R(\varphi(t), \psi(t))$$

on the trajectory, we get from the periodicity

$$\int_0^T \frac{1}{R}(P, Q) \cdot \operatorname{grad} R\, dt = [\log |R(\varphi(t), \psi(t))|]_0^T = 0\,.$$

Appendix B

Since $x = r\cos\theta$, $y = r\sin\theta$ it follows from (3) that

$$\begin{aligned} f_r'(\theta, r) &= \frac{\partial}{\partial r} \frac{xP + yQ}{-yP + xQ}\\ &= \frac{\left[\frac{x}{r}P + x\frac{\partial P}{\partial r} + \frac{y}{r}Q + y\frac{\partial Q}{\partial r}\right][-yP + xQ]}{[-yP + xQ]^2} - \frac{\left[-\frac{y}{r}P - y\frac{\partial P}{\partial r} + \frac{x}{r}Q + x\frac{\partial Q}{\partial r}\right][xP + yQ]}{[-yP + xQ]^2}\\ &= \frac{x^2 + y^2}{[-yP + xQ]^2}\left[Q\frac{\partial P}{\partial r} - P\frac{\partial Q}{\partial r}\right].\end{aligned}$$

Now $\frac{\partial P}{\partial r} = \frac{1}{r}\left[\frac{\partial P}{\partial x}x + \frac{\partial P}{\partial y}y\right]$, $\frac{\partial Q}{\partial r} = \frac{1}{r}\left[\frac{\partial Q}{\partial x}x + \frac{\partial Q}{\partial y}y\right]$ gives

$$r f_r'(\theta, r) = \frac{x^2 + y^2}{R(x, y)^2}\left[Q\left(x\frac{\partial P}{\partial x} + y\frac{\partial P}{\partial y}\right) - P\left(x\frac{\partial Q}{\partial x} + y\frac{\partial Q}{\partial y}\right)\right]. \tag{*}$$

From $R(x, y) = -yP(x, y) + xQ(x, y)$ it follows that

$$\frac{\partial R}{\partial x} = -y\frac{\partial P}{\partial x} + x\frac{\partial Q}{\partial x} + Q\,, \qquad \frac{\partial R}{\partial y} = -y\frac{\partial P}{\partial y} + x\frac{\partial Q}{\partial y} - P\,,$$

giving

$$y\frac{\partial P}{\partial y} = -\frac{\partial R}{\partial y} + x\frac{\partial Q}{\partial y} - P\,, \qquad x\frac{\partial Q}{\partial x} = \frac{\partial R}{\partial x} + y\frac{\partial P}{\partial x} - Q\,.$$

Hence

$$Q\left(x\frac{\partial P}{\partial x} + y\frac{\partial P}{\partial y}\right) - P\left(x\frac{\partial Q}{\partial x} + y\frac{\partial Q}{\partial y}\right) = Q\left(x\frac{\partial P}{\partial x} + x\frac{\partial Q}{\partial y} - \frac{\partial R}{\partial y} - P\right) - P\left(y\frac{\partial P}{\partial x} + y\frac{\partial Q}{\partial y} + \frac{\partial R}{\partial x} - Q\right)$$

$$= [-yP + xQ]\left(\frac{\partial P}{\partial x} + \frac{\partial Q}{\partial y}\right) - \left[P\frac{\partial R}{\partial x} + Q\frac{\partial R}{\partial y}\right] = R\,\mathrm{div}\,(P, Q) - (P, Q)\cdot\mathrm{grad}\,R$$

$$= R^2\left[\frac{1}{R}\,\mathrm{div}\,(P, Q) + (P, Q)\cdot\left(-\frac{1}{R^2}\,\mathrm{grad}\,R\right)\right] = R^2\,\mathrm{div}\,\frac{1}{R}\,(P, Q)\,.$$

Substituting this into the right hand side of (*) we get the formula (13).

KLEINE MITTEILUNGEN

ZAMM 60, 631 – 632 (1980)

M. A. Noor

Finite Element Error Analysis of Nonlinear Variational Inequalities

1. Introduction

Recently much attention has been paid to the application of the finite element method for obtaining numerical solutions of variational inequalities. Variational inequalities are closely related to partial differential equations whose solutions are required to satisfy some extra constraint conditions. For obstacle and unilateral problems involving linear variational inequalities, Falk [7], Mosco and Strang [9], Brezzi, Hager and Raviart [3], using piecewise linear elements, have shown that the error estimate for the finite element approximation converges with order $O(h)$.

In this paper, we analyze the error in the finite element approximation to nonlinear variational inequalities. We derive a general error estimate for the error in the approximation. We prove, using piecewise linear and conformal trial functions, that the error estimate in the energy norm is of order $O(h)$. The nonlinear problem is considered in Section 2. The error estimate is derived in Section 3.

2. The nonlinear obstacle problem

We consider the *nonlinear obstacle problem* of the type

$$\left.\begin{aligned}&\text{Find} \quad u \in K \quad \text{such that}\\ &(Tu, v - u) \geqq (f, v - u) \quad \text{for all } v \in K,\end{aligned}\right\} \tag{1}$$

where $f \in L_2(D)$ and K is defined by

$$K = \{v \in H^1;\ v \geqq \psi \text{ a.e. on } D,\ v = g \text{ on } S\},$$

$D \subset R^2$ is a bounded polygonal domain with boundary S, g and ψ are given functions on the boundary S and domain D, respectively.

Physical interpretation: In (1) u represents the small vertical displacement of an elastic membrane D under the effects of a field of vertical forces, whose intensity is given by f. This membrane is fixed on its boundary $S(u = g)$ and lies over an obstacle, whose height is given by $\psi(u \geqq \psi)$.

The space $W_2^k(D) = H^k$ is considered to be a usual Sobolev space, where for $u \in H^k$, k a positive integer, we have the norm

$$\|u\|_{k,D}^2 \equiv \sum_{|\alpha| \leqq k} \|D^\alpha u\|_{L_2(D)}^2 .$$

If T is a strongly monotone and Lipschitz continuous operator from $H^1(D)$ into $H^{-1}(D)$, being the dual space of $H^1(D)$, i.e. there exist constants $\varrho > 0$, $\mu > 0$ such that

$$(Tu - Tv, u - v) \geqq \varrho \|u - v\|^2 \quad \text{for all } u, v \in H^1$$

and

$$\|Tu - Tv\| \leqq \mu \|u - v\| \quad \text{for all } u, v \in H^1,$$

then there exists a unique solution of (1), cf. [10].

Furthermore, we assume that for the given functions ψ and $g \in H^2$, $u \in K$, satisfying (1), also is in H^2. Finally, because all the novelty (and difficulty) comes from ψ and not from f, we shall assume $f = 0$.

We need the following result for deriving the error estimate, which is due to Brezis [1].

Lemma 1: *Assume $u \in H^2$, then*

$$\left.\begin{aligned}&Tu \geqq 0 \quad \text{a.e. on } D,\\ &Tu(u - \psi) = 0 \quad \text{a.e. on } D.\end{aligned}\right\} \tag{2}$$

3. Error estimate for the finite element analysis

Let $\{T_h\}_{h>0}$ be a regular family (cf. [5]) of triangulation of D and define

$$S_h = \{v_h;\ v_h \in C^0(D),\ v_{h|T} \in P_1 \text{ for all } T \in T_h\},$$

where P_1 is the set of all polynomials on R^2 of degree $\leqq 1$. Clearly, S_h is a finite dimensional subspace of H^1. The convex set K_h is defined as

$$K_k = \{v_h;\ v_h \in S_h,\ v_h \geqq \psi \text{ at every vertex of the triangulation } T_h,\ v_h = g_h \text{ on } S\},$$

where g_h is the piecewise linear interpolant to g along the boundary S.

The *finite element approximation* of (2) is:

$$\begin{aligned}&\text{Find} \quad u_h \in K_h \quad \text{such that}\\ &(Tu_h, v_h - u_h) \geqq 0 \quad \text{for all } v_h \in K_h.\end{aligned} \tag{3}$$

For $K = H^1$ we have an equality instead of an inequality. Thus we have to construct only the finite dimensional subspace S_h. We mention that we do not require $K_h \subset K$. It is sufficient to consider the case where both K and K_h are in H^1. It is well known [11, 12] that for the equation $Tu = 0$ the error estimate depends on the inequality

$$\|u - u_h\| \leqq \text{constant} \cdot \text{dis}\,(u, S_h).$$

In general, this is not true for variational inequalities, cf. Mosco and Strang [9].

We now derive the *general error estimate* for $u - u_h$.

Theorem 1: *If u and u_h are the solutions of* (1) *and* (3), *then*

$$(Tu - Tu_h, u - u_h) \leqq (Tu - Tu_h, u - v_h) + (Tu, v_h - u) + (Tu, v - u_h) \quad \textit{for all } v \in K \textit{ and } v_h \in K_h. \tag{4}$$

Proof: Since u and u_h are solutions of (1) and (3), thus

$$(Tu, v - u) \geqq 0,$$

and

$$(Tu_h, v_h - u_h) \geqq 0.$$

Adding these inequalities and rearranging terms we get

$$(Tu, u) + (Tu_h, u_h) \leqq (Tu, v) + (Tu_h, v_h).$$

Subtracting from both sides $(Tu, u_h) + (Tu_h, u)$ we obtain

$$\begin{aligned}(Tu - Tu_h, u - u_h) &\leqq (Tu, v - u_h) + (Tu_h, v_h - u)\\ &= (Tu - Tu_h, u - v_h) + (Tu, v_h - u) + (Tu, v - u_h),\end{aligned}$$

which is (4), the required result.

Note that for $K = H^1$ the last two terms in (4) drop out and we have

$$(Tu - Tu_h, u - u_h) \leqq (Tu - Tu_h, u - v_h) \quad \text{for all } v_h \in S_h.$$

We remark that if T is both a strongly monotone and a Lipschitz continuous operator, then results

$$\|u - u_h\| \leqq \frac{\mu}{\varrho} \|u - v_h\| \quad \text{for all } v_h \in S_h,$$

a well known error bound (cf. [6]). For $u - u_h$ inequality (4) is a generalization of a result of Falk [7]. This fact allows us to consider nonlinear obstacle problems of type (1).

We now state and prove the main result of this paper.

Theorem 2: *Let $Tu \in L_2(D)$, $\psi \in H^2$, $g = g_{h|S} \in H^2$ and the angle condition be satisfied. If u and u_h are solutions of* (1) *and* (3) *then*

$$\|u - u_h\|_{1,D} = O(h). \tag{5}$$

Proof: Putting $v = u$ in inequality (4) we get

$$\begin{aligned}(Tu - Tu_h, u - u_h) &\leqq (Tu - Tu_h, u - v_h) + (Tu, u_h - u) + (Tu, u - u_h) + (Tu, v_h - u_h)\\ &= (Tu - Tu_h, u - v_h) + (Tu, v_h - u_h).\end{aligned} \tag{6}$$

Let I_h be the operator of S_h-interpolation. Then, since $D \subset R^2$, we have $H^2 \subset C^0(D)$, and $u \in H^2$, $u \geqq \psi$, on D imply $I_h u \in K_h$.

Taking $v_h = I_h u$ in (6) we obtain

$$(Tu - Tu_h, u - u_h) \leqq (Tu - Tu_h, u - I_h u) + (Tu, I_h u - u_h).$$

Consider now the term

$$(Tu, I_h u - u_h) = (Tu, I_h u - u) + (Tu, \psi - I_h \psi) + \\ + (Tu, u - \psi) + (Tu, I_h \psi - u_h) \leqq \\ \leqq (Tu, I_h u - u) + (Tu, \psi - I_h \psi); \quad (7)$$

then lemma 1 and the fact $u_h \in K_h$ imply $I_h \psi - u_h \leqq 0$ on D. Since $Tu \in L_2(D)$ it follows by making use of the CAUCHY-SCHWARTZ inequality that

$$(Tu, I_h u - u_h) \leqq ||Tu||_{L_2(D)} \{||I_h u - u||_{L_2(D)} + ||I_h \psi - \psi||_{L_2(D)}\} . \quad (8)$$

Since the angle condition and $u, \psi \in H^2$ hold, it follows that (cf. CIARLET [5])

$$||I_h u - u||_{r,D} \leqq C h^{2-r} ||u||_{2,D} , \qquad r = 0, 1 , \quad (9)$$

and

$$||I_h \psi - \psi||_{r,D} \leqq C h^{2-r} ||\psi||_{2,D} , \qquad r = 0, 1 , \quad (10)$$

where the C's are constants being independent of u, h and ψ.

Also, by using the LIPSCHITZ continuity of T we get

$$(Tu - Tu_h, I_h u - u) \leqq \tfrac{1}{2} ||u - u_h||^2_{1,D} + \tfrac{1}{2} C_1^2 h^2 ||u||^2_{2,D} . \quad (11)$$

Thus from (7)—(11) it follows that

$$||u - u_h||_{1,D} = O(h) , \quad (12)$$

which is the required estimate (5).

Remark: The problems of deriving the error estimates in L_2-norm and L_∞-norm for the nonlinear problems having constraint conditions are still open.

Acknowledgement

The author would like to thank the referee for his helpful suggestions, which influenced the final version of this paper.

References

1 BREZIS, H., Problèmes unilatéraux, J. Math. pures appl. 72 (1971), 1–68.
2 BREZIS, H.; STAMPACCHIA, G., Sur la régularité de la solution d'inéquations elliptiques, Bull. Soc. Math. France 96 (1968), 153–180.
3 BREZZI, F.; HAGER, W.; RAVIART, P. A., Error estimate for the finite element solution of variational inequalities, Numer. Math. 28 (1977), 431–443.
4 BREZZI, F.; SACCHI, G., A finite element approximation of a variational inequality related to hydraulics, Calcolo 13 (1976), 257–273.
5 CIARLET, P. G., The Finite Element Method for Elliptic Problems, North Holland, Amsterdam 1978.
6 CIARLET, P. G.; SCHULTZ, M., VARGA, R., Numerical methods of higher accuracy for nonlinear boundary value problems. IV, Numer. Math. 13 (1969), 51–77.
7 FALK, R. S., Error estimate for the approximation of a class of variational inequalities, Math. Comput. 28 (1974), 963–971.
8 GLOWINSKI, R., Introduction to the approximation of elliptic variational inequalities, T/R 76006, University of Paris VI, France, 1976.
9 MOSCO, U., STRANG, G., One-sided approximation and variational inequalities, Bull. Amer. Math. Soc. 80 (1974), 308–312
10 NOOR, M. A., On variational inequalities, Ph. D. Thesis, Brunel University, U.K. 1975.
11 STRANG, G.; FIX, G., An Analysis of the Finite Element Method, Prentice-Hall 1973.
12 ZLAMAL, M., Some recent advances in the mathematics of finite elements, in The Mathematics of Finite Elements and Applications, J. R. WHITEMAN (ed.), Academic Press, London-New York 1973.

Eingereicht am 2. 10. 1979, revidierte Fassung am 8. 1. 1980

Anschrift: Dr. M. ASLAM NOOR,
Department of Mathematics,
Islamia University, Bahawalpur, Pakistan

ZAMM 60, 632–634 (1980)

R. SUBRAMANIAN / N. RAVICHANDRAN

A Redundant System with Bulk Replacement

1. Introduction

Several models of standby redundant systems with repair and/or maintenance and/or replacement have been studied in the past. A bibliography of the work done so far has been prepared by OSAKI and NAKAGAWA [4], LIE, HWANG and TILLMAN [3] and PIERSKALLA and VOELKER [5]. It is clear from these that there are hardly any models dealing with bulk replacements.

This paper considers a *multi-component system with bulk replacement*. By identifying suitable regeneration points we obtain the *meantime to system failure* (MTSF) and *pointwise availability* of the system. The *stationary distribution* of *the underlying stochastic process* $\{X(t)\}$, representing the *number of operable units at any time* t, is also obtained. Using the expected cost per unit time in the stationary case, a method to determine the optimum number of spares is suggested.

2. Assumptions

1. The system requires $s'(>0)$ units for its successful operation.
2. Initially there are $S - s'$ (>0) spares available to replace the failed units.
3. When the number of operating units drops down to a level $s(\geqq s')$ an order for $(S - s)$ spares is placed. We assume that $(S - s) > s$.
4. The time interval between the placing of an order for the spares and its materialization is a random variable which is arbitrarily distributed. Its probability density function is denoted by $b(\cdot)$.
5. The failure rate of each of the operable units is assumed to be a constant λ, when the system is functioning and λ_1 when the system is down. We further assume that $\lambda_1 < \lambda$.

Thus the system considered is an s'-out-of-n:G system. Usually in the analysis of such systems it is assumed that when the system is down the remaining operable units do not fail. However, it is more reasonable and realistic to assume that when the system is down the operable units have constant hazard rate (See [8, 9]). Assumption 3 is essential. Otherwise it is possible that the number of operable units never reaches a level $\geqq s$; it might happen that before a replenishment occurs the number of operable units might drop down to zero and consequently the number of units after the replenishment will be $< s$. In such a situation further orderings can never take place.

3. Notations

$*$ denotes convolution; $c^{(n)}(t)$ is the n-fold convolution of an arbitrary function $c(t)$ with itself over $(0, t)$; $C(t) = \int_0^t c(u)\,du$; $\bar{c}(t) = 1 - c(t)$.

4. The stochastic process $\{X(t)\}$

The main object of interest is the stochastic process $\{X(t)\}$, a typical realisation of which is given in Figure 1. Note that when $X(t) = n$ and $n > s$, we can predict the future behaviour of the process without any additional information since no order is pending and the life times of the units are random variables with exponential distributions. However, when $X(t) = n$ and $n < s$, we require the time elapsed since the last reorder to predict the future behaviour of the process due to the non-exponential nature of the replenishment time. However, the process $\{X(t)\}$ has some regeneration points, the identification of which enables us to answer all the relevant questions regarding $\{X(t)\}$. One such set of points is formed by the epochs t_i at which $X(t_i) = s < X(t_i -)$. We define an α *event* as an event that corresponds to the dropping down of $X(t)$ to the value s. It is clear that the time intervals between successive α events are independently identically distributed random variables. The time interval between any two such α events will be called a *cycle* and its probability density function is denoted by

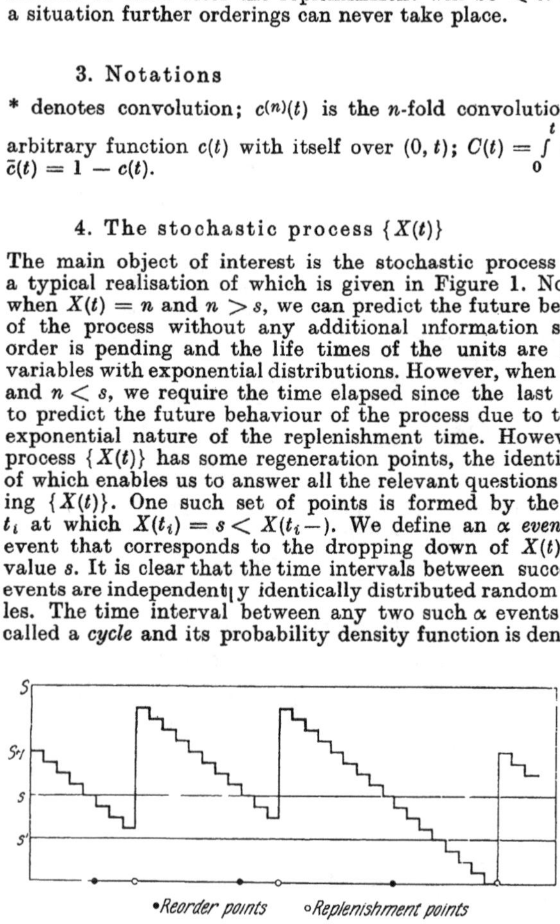

Fig. 1. Typical realisation of the stock inventory level

$f_\alpha(\cdot)$. The counting process associated with the α events will be denoted by $N_\alpha(t)$. We first analyse the process within a cycle, and use renewal theoretic arguments to obtain the distribution of $X(t)$ at any arbitrary time t. To carry out the analysis we also need the following function which is defined only for a replenishment time interval or a part thereof:

$$P_{ij}(t) = \mathrm{P}[X(t) = j \mid X(0) = i]\,, \qquad i, j = 0, 1, \dots, s\,.$$

We first prove

Theorem 1: *The probability density function of the time interval between two successive α events is given by*

$$f_\alpha(t) = \sum_{n=0}^{s} [b(t)\, P_{sn}(t)] * f_{S-s+n;\,s}(t) \tag{1}$$

where $f_{ij}(t) = \lambda_i\, \mathrm{e}^{-\lambda_i t} * f_{i-1,j}(t),\ i > j+1;\ f_{i+1,i}(t) = \lambda_{i+1}\, \mathrm{e}^{-\lambda_{i+1} t}$ *and* $f_{ii}(t) = 0$ *with*

$$\lambda_i = \begin{cases} i\lambda & \text{for } i \geqq s' \\ i\lambda_1 & \text{for } i < s'\,. \end{cases}$$

Proof: We note that for a reorder to occur in $(t, t + \mathrm{d}t)$, given that there is an order placed at $t \doteq 0$, it is necessary that the order placed at $t = 0$ materialises before t, say in $(u, u + \mathrm{d}u)$, $u < t$. This occurs with probability $b(u)\mathrm{d}u$. When the order materialises the system may or may not be operable and the number of operable units at this epoch is described by the functions $P_{ij}(t)$. Thus the number of operable units immediately after a replenishment is known. Using the functions $f_{..}(\cdot)$ the required probability is computed. Thus we get equation (1).

Next we determine the functions $P_{ij}(t)$. They are defined only for a replenishment time interval or a part thereof and further only for $j \leqq i$. By identifying the behaviour of the system in a replenishment time interval with a state dependent death process we get

$$P_{ij}(t) = f_{ij}(t) * \mathrm{e}^{-\lambda_j t} + \delta_{ij}\, \mathrm{e}^{-\lambda_j t}\,.$$

Next we have

Theorem 2: *Let*

$\Psi(n, t) = \mathrm{P}[X(t) = n;\ N_\alpha(t) - N_\alpha(0) = 0 \mid \alpha \text{ event at } t = 0]\,.$

The functions $\Psi(n, t)$, $0 \leqq n \leqq S$, are given by

$$\Psi(n, t) = \bar{B}(t)\, P_{sn}(t)\,, \qquad 0 \leqq n \leqq s\,, \tag{2}$$

$$\Psi(n, t) = \sum_{n'=0}^{s} [h(t)\, P_{sn'}(t)] * P_{n'+S-s;\,n}(t)\,, \qquad s < n \leqq S \tag{3}$$

where

$$h(t) = \sum_{n=1}^{\infty} f_\alpha^{(n)}(t)\,, \tag{4}$$

$f_\alpha(\cdot)$ being given by equation (1).

The proof of the theorem is straight forward. Next we prove

Theorem 3: *The functions $\pi(n, t)$ $(0 \leqq n \leqq S)$ representing the probability mass function of $X(t)$ at time t, conditioned by an α event at the origin is given by*

$$\pi(n, t) = \Psi(n, t) + h(t) * \Psi(n, t)\,, \qquad 0 \leqq n \leqq S\,,$$

where $h(t)$ is given by equation (4).

The theorem follows by noting the fact that the level can reach n either in the first cycle or in a subsequent cycle.

It is also possible to deduce the stationary distribution of $X(t)$ from Theorem 3. Setting $\pi(n) = \operatorname{Lt}_{t\to\infty} \pi(n, t)$, we get from Theorem 3, as direct consequence of the key renewal theorem [7]

$$\pi(n) = \int_0^\infty \Psi(n, t)\, \mathrm{d}t \Big/ \int_0^\infty t f_\alpha(t)\, \mathrm{d}t\,.$$

5. The operating characteristics of the system

Using the results obtained in the previous sections we can discuss the operating characteristics of the system such as the reliability and availability of the system. The availability of the system conditioned at the time origin by an α event is given by $\sum_{n=s'}^{S} \pi(n, t)$ and the same in the stationary case is $\sum_{n=s'}^{S} \pi(n)$.

To obtain the reliability of the system we need certain new functions which can be easily obtained from the functions introduced earlier. We use a "$\sim$" to denote the new function obtained from a given function. The new functions are defined only over an interval in which there is no system failure. For example $\pi(n, t)$ denotes the probability that $X(t) = n$, given that at $t = 0$ there is an α event. Then $\tilde{\pi}(n, t)$ will denote the probability that $X(t) = n$ with same initial conditions and in addition $X(u) \geqq s'$ for $0 \leqq u \leqq t$. With this modification the structure of equations (1)—(5) remains unchanged. However in equation (1) only the first term survives. Thus we get $\tilde{\pi}(n, t)$ for $s' \leqq n \leqq S$. The reliability of the system is given by $\sum_{n=s'}^{S} \tilde{\pi}(n, t)$ and the MTSF is given by $\sum_{n=s'}^{S} \tilde{\pi}^*(n)$ where $\tilde{\pi}^*(n) = \int_0^\infty \tilde{\pi}(n, t)\, \mathrm{d}t$.

Using the above results, we proceed to set up an expression for the expected cost per unit time in the stationary case. We introduce the following costs.

C_v — unit variable cost,
C_f — fixed cost for replacement,
p — penalty cost per unit per unit time when the system is down,
C — storage cost per unit per unit time.

Then the *expected cost per unit time* is seen to be

$$C = \frac{C_f + (S - s)\, C_v}{M} + C_s \sum_{n=s'}^{S} (n - s')\, \pi(n) + p \sum_{n=0}^{s'-1} \pi(n)$$

where $M = \int_0^\infty t f_\alpha(t)\, \mathrm{d}t$.

6. The associated inventory problem

The stochastic process $\{X(t)\}$ can also be seen to describe an inventory problem. It corresponds to an inventory model with an ordering policy of the (S, s) type in which the demand is state dependent and lead time is a random variable with an arbitrary distribution. Then $X(t)$ will denote the inventory level at any time t. A typical realization of the inventory level is exhibited in Figure 1. This inventory problem has a peculiar character; when the stock level is zero demands do not arise temporarily till the next replenishment, in contrast with the usual inventory problem [1, 2]. The functions $\pi(n, t)$ and $\pi(n)$ describe the inventory level in the transient and stationary cases, respectively.

7. Particular cases

We now discuss the special case where the lead time is a constant equal to t_0 (i.e.) $b(t) = \delta(t - t_0)$. The equations representing various probabilities are given by

$$f_\alpha(t) = \sum_{n=0}^{S} P_{sn}(t_0)\, f_{S-s+n;\,s}(t - t_0)\,,$$

$$\Psi(n, t) = \begin{cases} P_{sn}(t) & t < t_0 \\ 0 & t \geqq t_0, \end{cases} \qquad 0 \leqq n \leqq s\,,$$

$$\Psi(n, t) = \sum_{n'=0}^{s} P_{sn'}(t_0)\, P_{n'+S-s;\,n}(t - t_0) \qquad s < n \leqq S\,.$$

The case $t_0 = 0$ is discussed in [6]. Setting $t_0 = 0$ we get,

$$\pi(n) = 0\,, \qquad 0 < n \leqq s\,, \qquad \pi(n) = 1 \Big/ \left(n \sum_{i=s+1}^{s} \frac{1}{i}\right),$$

a result in agreement with Sivazlian and Stanfel [6, p. 319].

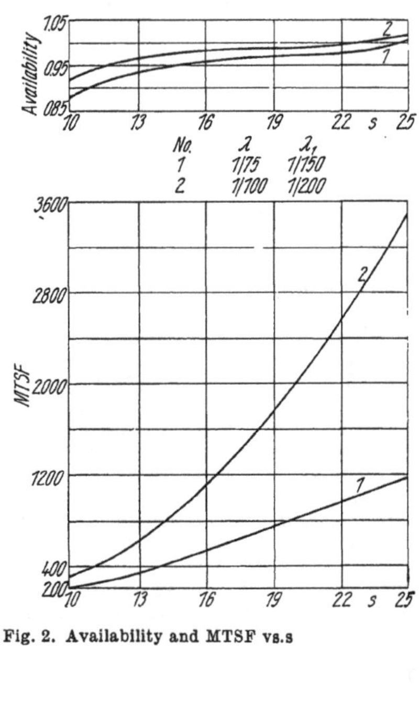

Fig. 2. Availability and MTSF vs. s

8. Numerical results

We assume that the system requires at least 10 units for functioning and the maximum number that the system can accommodate is 50 i.e. $s' = 10$ and $S = 50$. We also assume the following values for the other parameters. The results are shown graphically in Figures 2 and 3.

$$C_f = 20\,, \quad p = 100\,, \quad C_s = 1\,, \quad C_v = 10\,,$$

$$\lambda = 1/50\,, \quad b(t) = b\,e^{-bt} : b = 1/25\,, \quad \lambda_1 = 1/100,\ 1/200\,.$$

From the figures we find that $s = 16$ and $s = 14$ are the optimal values at which the reorder is to be placed. The availability and MTSF are increasing functions of s, as is to be expected.

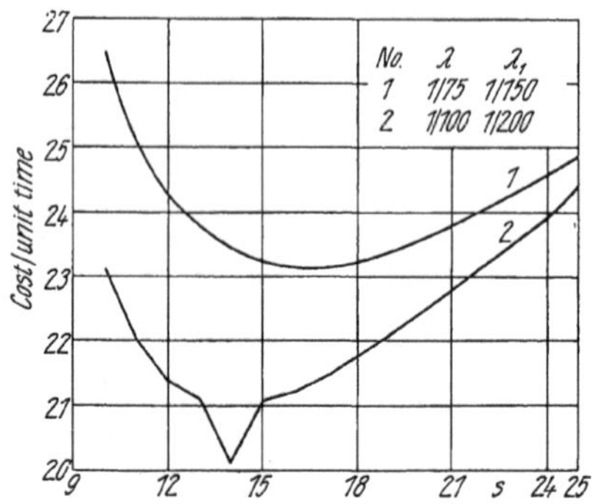

Fig. 3. Cost per unit time vs.s

Acknowledgement

One of the authors (N. RAVICHANDRAN) is thankful to CSIR for financial support during the preparation of this manuscript.

References

1 ARROW, K. J., KARLIN S.; SCARF, H., Studies in the Mathematical Theory of Inventory and Production, Standford University Press 1958

2 HADLEY, G.; WHITIN, T. M., Analysis of Inventory Systems, Prentice Hall Inc., Eglewood Cliffs, New Jersey 1962

3 LIE, D. H.; HWANG, C. L.; TILLMAN, F. A., Availability of maintained systems: A state of the art survey, AIIE Trans. **9**, 247–259 (1977)

4 OSAKI, S.; NAKAGAWA, T., Bibliography for reliability and availability of stochastic systems, IEEE Trans. Rel. **R-25**, 284–287 (1976)

5 PIERSKA, W. P.; VOELKER, J. A., A survey of maintenance models: the control of surveillance of deteriorating systems, Naval Res. Logist. Quart. **23**, 353–388 (1976)

6 SIVAZLIAN, B. D.; STANFEL, L. E., Analysis of Systems in Operations Research, Prentice Hall Inc., Englewood Cliffs, New Jersey 1975

7 SRINIVASAN, S. K., Stochastic Point Processes and Their Applications, Griffin, London 1974

8 SUBRAMANIAN, R., Availability of a GNEDENKO system, IEEE Trans. Rel. **R-62**, 302–303 (1977)

9 SUBRAMANIAN, R.; RAVICHANDRAN, N., Availability of a redundant system, IEEE Trans. Rel. **R-27**, 237–238 (1978)

Eingereicht: 19. 11. 1979, korrigierte Fassung: 8. 5. 1980

Anschrift: Dr. R. SUBRAMANIAN; N. RAVICHANDRAN, Indian Institute of Technology, Department of Mathematics, Madras-600036, India

ZAMM 60, 634 - 635 (1980)

H. PREDIGER

Aufbereitung von Volumenintegralen in der Starrkörpermechanik zur programmierten Berechnung

Bei der kinetischen Untersuchung von Konstruktions- und Bauelementen mit Mitteln der Starrkörpermechanik werden als Vorwerte die Volumenintegrale *Massenträgheitstensor* (Abk. MTT) und Volumen benötigt. Ein weites Anwendungsfeld ist die Ermittlung des Schwingungsverhaltens von Maschinenblockfundamenten, s. [3]. Eine Aufbereitung der Integrationsaufgabe für den Computer ist zur Vermeidung von Rechenfehlern und zur Rationalisierung des Arbeitsaufwandes wünschenswert. Voraussetzung zur Durchführung des hier vorgeschlagenen Verfahrens ist, daß die Dichte des Körpers homogen ist. Dann ist die Aufgabe über die Oberfläche zu lösen analog einem in [2] angegebenen Verfahren für Flächenintegrale. Es seien 0 ein beliebiger Bezugspunkt und $\boldsymbol{x}$ der 0 zugeordnete Ortsvektor. Dann sind das *statische Moment* $\boldsymbol{S}$ und der *auf die Dichte* ϱ *bezogene* MTT Θ bezüglich 0 definiert durch die Volumenintegrale

$$\boldsymbol{S} = \int_V \boldsymbol{x}\, dv\,, \tag{1}$$

$$\Theta = \int_V (\boldsymbol{x} \cdot \boldsymbol{x}\, \mathbf{1} - \boldsymbol{x} \otimes \boldsymbol{x})\, dv\,. \tag{2}$$

Die Tensornotation hält sich an [1]. Zur Umwandlung der Volumenintegrale in Oberflächenintegrale wird auf die Integralsätze von GAUSS zurückgegriffen

$$\int_V \operatorname{div} \boldsymbol{u}\, dv = \int_A \boldsymbol{u} \cdot d\boldsymbol{a}; \qquad \int_V \operatorname{div} \boldsymbol{T}\, dv = \int_A \boldsymbol{T}\, d\boldsymbol{a}\,, \tag{3 a, b}$$

worin $\boldsymbol{u}$ ein beliebiges Vektor- und $\boldsymbol{T}$ ein Tensorfeld sind. Setzt man in (3a) $\boldsymbol{u} = \boldsymbol{x}$, wird der Integrand der linken Seite 3, und das Volumen V stellt sich als das Oberflächenintegral

$$V = \frac{1}{3} \int_A \boldsymbol{x} \cdot d\boldsymbol{a} \tag{4}$$

dar. Wählt man in (3b) $\boldsymbol{T} = \boldsymbol{x} \otimes \boldsymbol{x}$, so schließt man im Vergleich mit (1), daß das statische Moment sich als

$$\boldsymbol{S} = \frac{1}{4} \int_A (\boldsymbol{x} \otimes \boldsymbol{x})\, d\boldsymbol{a} = \frac{1}{4} \int_A \boldsymbol{x}(\boldsymbol{x} \cdot d\boldsymbol{a}) \tag{5}$$

schreiben läßt. Das Integral in (2) läßt sich nicht in geschlossener Form umwandeln. Wir errichten in 0 ein kartesisches Basissystem $\boldsymbol{e}_i$ und ermitteln den MTT komponentenweise. Für die Diagonalkomponente Θ_{11} geben wir in (3a)

$$u_1 = \frac{1}{3} (x_2)^3; \qquad u_2 = \frac{1}{3} (x_3)^3; \qquad u_3 = 0 \tag{6}$$

vor und schließen im Vergleich mit der Komponentenform von (2) auf

$$\Theta_{11} = \frac{1}{3} \int_A [(x_2)^3\, da_2 + (x_3)^3\, da_3]\,. \tag{7}$$

Andere Ansätze als (6) führen ebenfalls zum Ziel. Durch zyklische Vertauschung der Indizes in (7) gelangt man zu Darstellungen für Θ_{22} bzw. Θ_{33}. Für die Deviationsmomente Θ_{ik} $(i \neq k)$ setzen wir in (3a) die Komponenten

$$u_i = 0; \qquad u_{i+1} = -\frac{1}{2} x_i (x_{i+1})^2; \qquad u_{i+2} = 0 \tag{8}$$

und gewinnen aus dem Vergleich mit (2)

$$\Theta_{ik} = -\frac{1}{2} \int_A x_i (x_{i+1})^2\, da_{i+1}; \qquad i \neq k\,, \tag{9}$$

wobei zu beachten ist, daß doppelt auftretende Indizes keine Summationsvorschrift im Sinne der Tensorrechnung anzeigen.

Wir spezifizieren die Ergebnisse nun für solche *Körper, die durch ebene Begrenzungsflächen berandet sind.* Da ihre Konfigurationen durch die Koordinaten der Eckpunkte bestimmt sind, eignen sie sich insbesondere für die Berechnung durch Programme, wobei gekrümmte Oberflächen durch Festlegung geeigneter Ersatzpunkte beliebig approximiert werden können. Zur eindeutigen Beschreibung der Geometrie ist allerdings eine bestimmte Systematik einzuhalten, wie sie in Abb. 1 skizziert ist. Die Teiloberflächen sind durch Randpolygonzüge eingefaßt. Die Gestalt des Körpers kann widerspruchsfrei durch ein Programm identifiziert werden, wenn die Koordinaten der n Polygonpunkte jeder m-ten Teilfläche im gleichen Richtungssinn

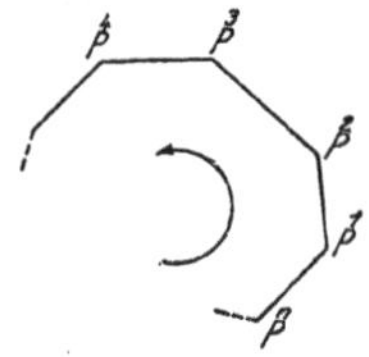

Bild 1. Außenseite einer ebenen Teilfläche

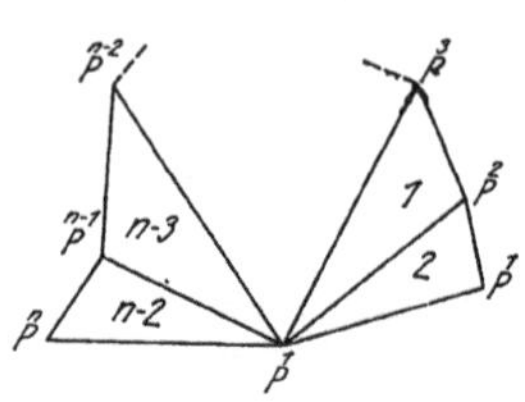

Bild 2. Einteilung der Teilfläche in $n - 2$ Dreiecke (In der Bezeichnung der Flächen sind 1 und 2 zu vertauschen)

eingegeben werden. Die Auswertung der Oberflächenintegrale (4), (5), (7) und (9) wird zunächst für eine Dreieckfläche vorgenommen, anschließend kann automatisch durch das Programm oder durch Erweiterung der gewonnenen Summenformeln die Summation für die $n - 2$ Dreiecke (Abb. 2) durchgeführt werden. Die Dreieckfläche 1 (Abb. 2) habe die Eckpunkte $\overset{1}{P}$, $\overset{2}{P}$ und $\overset{3}{P}$ mit den Ortsvektoren $\overset{1}{\boldsymbol{x}}$, $\overset{2}{\boldsymbol{x}}$ und $\overset{3}{\boldsymbol{x}}$. Eine Vektorfunktion der Fläche in Abhängigkeit der Eckpunkte kann sinnvoll mit den GAUSSschen Parametern t^1 und t^2 durch

$$\boldsymbol{x}(t^1, t^2) = \overset{1}{\boldsymbol{x}} + t^1(\overset{2}{\boldsymbol{x}} - \overset{1}{\boldsymbol{x}}) + t^1 t^2(\overset{3}{\boldsymbol{x}} - \overset{2}{\boldsymbol{x}}) \tag{10}$$

aufgestellt werden, wobei die Parameter die Wertebereiche

$$0 \leqq t^1 \leqq 1; \qquad 0 \leqq t^2 \leqq 1 \tag{11}$$

belegen. Die natürlichen Basisvektoren berechnen sich zu

$$\begin{aligned} \boldsymbol{a}_1 &= \boldsymbol{x}_{,1} = \overset{2}{\boldsymbol{x}} - \overset{1}{\boldsymbol{x}} + t^2(\overset{3}{\boldsymbol{x}} - \overset{2}{\boldsymbol{x}}), \\ \boldsymbol{a}_2 &= \boldsymbol{x}_{,2} = t^1(\overset{3}{\boldsymbol{x}} - \overset{2}{\boldsymbol{x}}). \end{aligned} \tag{12}$$

Der mathematisch positive differentielle Flächenvektor $\mathrm{d}\boldsymbol{a}$ wird aus dem äußeren Produkt $\mathrm{d}t^1\boldsymbol{a}_1 \times \mathrm{d}t^2\boldsymbol{a}_2$ gebildet und ergibt sich zu

$$\mathrm{d}\boldsymbol{a} = \sum_{i=1}^{3} (\overset{i}{\boldsymbol{x}} \times \overset{i+1}{\boldsymbol{x}})\, t^1\, \mathrm{d}t^1\, \mathrm{d}t^2 \qquad (i \text{ zyklisch}). \tag{13}$$

Im folgenden werden die Teilintegrale über die Dreieckflächen mit $\Delta(\ldots)$ bezeichnet. Nützlich für die praktische Bearbeitung ist die Berechnung des Flächenvektors, denn wegen des „Flächensatzes" $\int_A \mathrm{d}\boldsymbol{a} = 0$ für die geschlossene Oberfläche läßt sich die Forderung

$$\sum_{j=1}^{m} (\Delta\boldsymbol{A})_j = 0 \tag{14}$$

zur Kontrolle der Eingabedaten verwenden. Die Integration von (13) über die Dreieckfläche führt unter Beachtung von (11) auf

$$\Delta\boldsymbol{A} = \int_{\Delta A} \mathrm{d}\boldsymbol{a} = \tfrac{1}{2} \sum_{i=1}^{3} \overset{i}{\boldsymbol{x}} \times \overset{i+1}{\boldsymbol{x}}. \tag{15}$$

Weiterhin liefert die Integration von (4) und (5)

$$\Delta V = \tfrac{1}{6}\, \overset{1}{\boldsymbol{x}} \cdot (\overset{2}{\boldsymbol{x}} \times \overset{3}{\boldsymbol{x}}), \tag{16}$$

$$\Delta\boldsymbol{S} = \tfrac{1}{4}\, (\overset{1}{\boldsymbol{x}} + \overset{2}{\boldsymbol{x}} + \overset{3}{\boldsymbol{x}})\, \Delta V. \tag{17}$$

Die Integrationsergebnisse für den MTT werden wieder getrennt für die *Diagonalelemente*

$$\begin{aligned} \Delta\Theta_{(kk)} &= \tfrac{1}{30} \sum_{i=k+1}^{k+2} \Big(\overset{1}{x}_i\overset{1}{x}_i\overset{1}{x}_i + \overset{1}{x}_i\overset{1}{x}_i\overset{2}{x}_i + \overset{1}{x}_i\overset{1}{x}_i\overset{3}{x}_i + \overset{1}{x}_i\overset{2}{x}_i\overset{2}{x}_i + \overset{1}{x}_i\overset{2}{x}_i\overset{3}{x}_i + \\ &\quad + \overset{1}{x}_i\overset{3}{x}_i\overset{3}{x}_i + \overset{2}{x}_i\overset{2}{x}_i\overset{2}{x}_i + \overset{2}{x}_i\overset{2}{x}_i\overset{3}{x}_i + \overset{2}{x}_i\overset{3}{x}_i\overset{3}{x}_i + \overset{3}{x}_i\overset{3}{x}_i\overset{3}{x}_i\Big) (\Delta\boldsymbol{A})_i \end{aligned} \tag{18}$$

und die *Deviationsmomente*

$$\begin{aligned} \Delta\Theta_{ik} = -\tfrac{1}{60}\Big[&\overset{1}{x}_i\Big(3\overset{1}{x}_k\overset{1}{x}_k + 2\overset{1}{x}_k\overset{2}{x}_k + 2\overset{1}{x}_k\overset{3}{x}_k + \overset{2}{x}_k\overset{2}{x}_k + \overset{2}{x}_k\overset{3}{x}_k + \\ &+ \overset{3}{x}_k\overset{3}{x}_k\Big) + \overset{2}{x}_i\Big(\overset{1}{x}_k\overset{1}{x}_k + 2\overset{1}{x}_k\overset{2}{x}_k + \overset{1}{x}_k\overset{3}{x}_k + 3\overset{2}{x}_k\overset{2}{x}_k + 2\overset{2}{x}_k\overset{3}{x}_k + \\ &+ \overset{3}{x}_k\overset{3}{x}_k\Big) + \overset{3}{x}_i\Big(\overset{1}{x}_k\overset{1}{x}_k + \overset{1}{x}_k\overset{2}{x}_k + 2\overset{1}{x}_k\overset{3}{x}_k + \overset{2}{x}_k\overset{2}{x}_k + 2\overset{2}{x}_k\overset{3}{x}_k + \\ &+ 3\overset{3}{x}_k\overset{3}{x}_k\Big)\Big] (\Delta\boldsymbol{A})_k \\ &(\text{für } i \neq k) \end{aligned} \tag{19}$$

mitgeteilt. Summation der Resultate für die Teil- und Gesamtoberfläche schließen die Berechnung für den Bezugspunkt 0 ab. Die Transformation auf die Schwerpunkthauptachsen liefert schließlich die Vorgabewerte zur Bearbeitung des kinetischen Problems.

Zusammenfassung

Zur Ermittlung von Volumen und Massenträgheitstensor von Körpern homogener Dichte werden Formeln angegeben, welche sich zur Bearbeitung durch ein Rechenprogramm eignen. Dabei wird vorausgesetzt, daß die Körperoberfläche bereichsweise aus Ebenen gebildet wird.

Literatur

1 DE BOER, R.; PREDIGER, H., Tensorrechnung, Forschungsberichte aus dem Fachbereich Bauwesen, Universität Essen GH, 1978.
2 FLESSNER, H., Ein Beitrag zur Ermittlung von Querschnittswerten mit Hilfe elektronischer Rechenanlagen, Der Bauingenieur (1962), 146–149.
3 MÜLLER, F. P., Baudynamik, Betonkalender 1978 II, S. 745ff.

Anschrift: Dipl.-Ing. H. PREDIGER, Universität Essen GH, Fachbereich 10, Universitätsstraße 15, D-4300 Essen, BRD

ZAMM 60, 635–636 (1980)

NIRMAL C. SACHETI

On the Steady State Temperature Distribution in the Thermal Boundary Layer of an Elastico-Viscous Fluid

1. Introduction

In a short note, POP and SOUNDALGEKAR [1] have recently discussed the solution of the thermal boundary layer in the steady flow of WALTERS liquid B'' [2] parallel to an infinite porous plate with uniform suction. The analysis has been carried out neglecting elastic dissipation effects and using the velocity expression due to FRATER [3].

The present note aims to reconsider the above problem in the belief that elastic dissipation effects in the thermal boundary layer region are quite important for an accurate qualitative picture of non-NEWTONian behaviour in relation to NEWTONian fluid behaviour. Accordingly, the modified energy equation has been solved using the same method. The conclusions drawn from the present analysis clearly contrast the findings of POP and SOUNDALGEKAR [1].

2. Analysis

On including the elastic dissipation terms, the equation (7) of POP and SOUNDALGEKAR [1] modifies to

$$-c_p v_0 \frac{\mathrm{d}T}{\mathrm{d}y} = \frac{\lambda}{\varrho} \frac{\mathrm{d}^2T}{\mathrm{d}y^2} + \nu \left(\frac{\mathrm{d}u}{\mathrm{d}y}\right)^2 + \frac{K_0}{\varrho} v_0 \frac{\mathrm{d}u}{\mathrm{d}y} \frac{\mathrm{d}^2u}{\mathrm{d}y^2}. \tag{1}$$

Now introducing the non-dimensional quantities [1], the corresponding equations (10) and (11) become

$$Kf''' + f'' + f' = 0, \tag{2}$$

$$\theta'' + \sigma\theta' = -E\sigma(f'^2 + Kf'f''), \tag{3}$$

with boundary conditions

$$\begin{aligned} &f = 0, \quad \frac{\mathrm{d}\theta}{\mathrm{d}\eta} = 0 \quad \text{at} \quad \eta = 0, \\ &f \to 1, \quad \theta \to 0 \quad \text{as} \quad \eta \to \infty. \end{aligned} \tag{4}$$

The solution of (3), subject to FRATER's solution [3]

$$f(\eta) = 1 - \mathrm{e}^{-(1+K)\eta}$$

has been obtained as

$$\theta(\eta) = \frac{E(1 - 2K^2 - K^3)}{2K - \sigma + 2} \left[(1 + K)\, \mathrm{e}^{-\sigma\eta} - \frac{\sigma}{2}\, \mathrm{e}^{-2(1+K)\eta}\right], \tag{5}$$

which is not the same as the corresponding equation (14) of POP and SOUNDALGEKAR [1] in that (5) has a different factor outside the square bracket due to inclusion of elastic dissipation terms.

3. Conclusions

Figures 1 and 2 show the temperature distribution in the non-NEWTONian fluids with and without inclusion of elastic dissipation effects, along with the corresponding NEWTONian temperature distribution for a comparative study. It may be concluded that the temperature in the boundary layer for elastico-viscous fluids under consideration is always less than the corresponding NEWTONian fluid temperature ($K = 0$) for all values of the PRANDTL number (σ). This is clearly not in agreement with the findings of POP and SOUNDALGEKAR. It may finally be mentioned that the results obtained in [1] for $K = 0.3$ may questionable in view of FRATER's analysis [3].

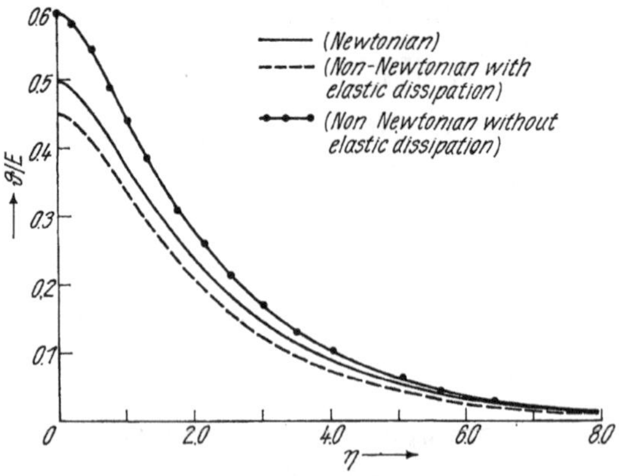

Fig. 1. Temperature distribution in the boundary layer for PRANDTL number $\sigma = 0.5$ and $K = 0$ (NEWTONian), $K = 0.2$ (Non-NEWTONian)

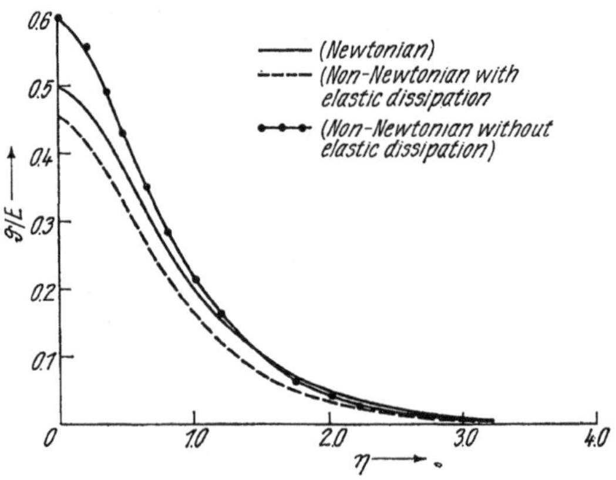

Fig. 2 Temperature distribution in the boundary layer for PRANDTL number $\sigma = 2.0$ and $K = 0$ (NEWTONian), $K = 0.2$ (Non-NEWTONian).

References

1 POP, I. and SOUNDALGEKAR, V. M., ZAMM 57 (1977) 493.
2 WALTERS, K., J. Mecan. 1 (1962) 479.
3 FRATER, K. R., ZAMP 21 (1970) 134.

Eingereicht am 15. 10. 1979

Anschrift: Dr. NIRMAL C. SACHETI, Department of Mathematics, Monash University, Clayton, Victoria 3168, Australia.

ZAMM 60, 636 – 637 (1980)

D. HOMENTCOVSCHI

On the Electrostatic Potential Problem for a Spherical Cap

The electrostatic potential due to a spherical cap, maintained at a given potential b, in an external electrical field is determined. We use spherical coordinates (r, θ, φ) refered to the centre of the sphere as origin, and the axis of the cap — as polar axis. In these coordinates the *equation of the spherical cap* is

$$r = a; \qquad 0 \leqq \theta \leqq \alpha; \qquad 0 \leqq \varphi < 2\pi .$$

The *external electrical field* is characterised by the potential $V_0(r, \theta, \varphi)$. This is a harmonic function in a neighbourhood of the body and consequently on the cap it can be written as the FOURIER series

$$V_0(a, \theta, \varphi) = - \sum_{n=0}^{\infty} f_n(\theta) \cos n(\varphi - \varphi_n) , \qquad 0 \leqq \theta \leqq \alpha; \qquad 0 \leqq \varphi < 2\pi . \tag{1}$$

Here $f_n(\theta)$ are continuously differentiable functions and φ_1, φ_2, ... are constants.

A similar problem in the case of the electrified disk was solved by COMPSON [1]. COLLINS [2] considered the problem of the electrified spherical cap in the case of the external field symmetrical about the axis of the body.

Let

$$g(\theta, \varphi) = \sum_{n=0}^{\infty} g_n(\theta) \cos n(\varphi - \varphi_n) , \qquad 0 \leqq \theta \leqq \alpha; \quad 0 \leqq \varphi < 2\pi,$$

be the *charge density* induced on the cap. The *potential at the point* (r, θ, φ) due to this charge density is

$$V_b(r, \theta, \varphi) = \int_0^{\alpha} \int_0^{2\pi} \frac{\sum_{n=0}^{\infty} g_n(\theta') \cos n(\varphi' - \varphi_n)\, a^2 \sin \theta' \, d\theta' \, d\varphi'}{\sqrt{r^2 + a^2 - 2ar\,(\cos \theta \cos \theta' + \sin \theta \sin \theta' \cos (\varphi' - \varphi))}} .$$

By setting $\varphi' - \varphi = \psi$ we get

$$V_b(r, \theta, \varphi) = \sum_{n=0}^{\infty} \cos n(\varphi - \varphi_n) \cdot \int_0^{\alpha} g_n(\theta') \sin \theta' \, d\theta' \int_0^{2\pi} \frac{a^2 \cos n\psi \, d\psi}{\sqrt{r^2 + a^2 - 2ar \cos \gamma}} ,$$

where $\cos \gamma = \cos \theta \cos \theta' + \sin \theta \sin \theta' \cos \psi$. We shall transform the last integral in this formula by setting $e^{i\psi} = \mathfrak{z}$

$$\int_0^{2\pi} \frac{a^2 \cos n\psi \, d\psi}{\sqrt{r^2 + a^2 - 2ar \cos \gamma}} = - \frac{a^2}{\sqrt{ar \sin \theta \sin \theta'}} \int_{|\mathfrak{z}|=1} \frac{\mathfrak{z}^n \, d\mathfrak{z}}{\sqrt{\mathfrak{z}(\mathfrak{z} - \xi_1)(\mathfrak{z} - \xi_2)}} , \tag{2}$$

where

$$\xi_{1,2} = \frac{r^2 + a^2 - 2ar \cos \theta \cos \theta'}{2ar \sin \theta \sin \theta'} \mp \frac{[(r^2 + a^2 - 2ar \cos (\theta - \theta'))\,(r^2 + a^2 - 2ar \cos (\theta + \theta'))]^{1/2}}{2ar \sin \theta \sin \theta'}$$

$(\xi_1 < \xi_2)$.

The square root in (2) is a holomorphic function in the whole plane with the cuts $[0, \xi_1]$, $[\xi_2, +\infty)$ on the positive real axis. Therefore, we can deform the integration contour such that it coincides with the cut $[0, \xi_1]$. We have

$$\int_0^{2\pi} \frac{a^2 \cos n\psi \, d\psi}{\sqrt{r^2 + a^2 - 2ar \cos \gamma}} = - \frac{2a^2}{\sqrt{ar \sin \theta \sin \theta'}} \int_0^{\xi_1} \frac{\xi^n \, d\xi}{\sqrt{\xi(\xi_1 - \xi)(\xi_2 - \xi)}} . \tag{3}$$

Hence

$$V_b(r, \theta, \varphi) = \sum_{n=0}^{\infty} \cos n(\varphi - \varphi_n) \frac{2a^2}{\sqrt{ar \sin \theta}} \times \int_0^{\alpha} \frac{g_n(\theta') \sin \theta' \, d\theta'}{\sqrt{\sin \theta'}} \int_0^{\xi_1} \frac{\xi^n \, d\xi}{\sqrt{\xi(\xi_1 - \xi)(\xi_2 - \xi)}} . \tag{4}$$

The boundary condition requires that the *total potential* $V = V_0 + V_b$ on the spherical cap *is constant*

$$V_b(a, \theta, \varphi) + V_0(a, \theta, \varphi) = b , \qquad 0 \leqq \theta \leqq \alpha; \qquad 0 \leqq \varphi < 2\pi . \tag{5}$$

By using the expressions (1) and (4) of the potentials in relation (5) we obtain the integral equation for the determination of the FOURIER coefficient $g_n(\theta)$

$$\int_0^{\alpha} \frac{g_n(\theta') \, a \sin \theta' \, d\theta'}{\sqrt{\sin \theta \sin \theta'}} \int_0^{\tilde{\xi}_1} \frac{2\xi^n \, d\xi}{\sqrt{\xi(\tilde{\xi}_1 - \xi)(\tilde{\xi}_2 - \xi)}} = b\delta_{n0} + f_n(\theta) , \tag{6}$$

where

$$\tilde{\xi}_1 = \min\left\{\frac{\operatorname{tg}\frac{\theta}{2}}{\operatorname{tg}\frac{\theta'}{2}} ; \frac{\operatorname{tg}\frac{\theta'}{2}}{\operatorname{tg}\frac{\theta}{2}}\right\} ; \tilde{\xi}_2 = \max\left\{\frac{\operatorname{tg}\frac{\theta}{2}}{\operatorname{tg}\frac{\theta'}{2}} ; \frac{\operatorname{tg}\frac{\theta'}{2}}{\operatorname{tg}\frac{\theta}{2}}\right\}.$$

By the change of the variable $\xi \operatorname{tg}\frac{\theta}{2}\cdot\operatorname{tg}\frac{\theta'}{2} = \operatorname{tg}^2\frac{\theta_1}{2}$, $(\xi \to \theta_1)$ the equation (6) becomes

$$\int_0^\alpha \frac{2g_n(\theta')\, a \sin\theta'}{\operatorname{tg}^n\frac{\theta}{2}\operatorname{tg}^n\frac{\theta'}{2}}\, d\theta' \int_0^{\min(\theta,\theta')} \frac{\operatorname{tg}^{2n}\frac{\theta_1}{2}\, d\theta_1}{\sqrt{(\cos\theta_1 - \cos\theta)(\cos\theta_1 - \cos\theta')}} =$$

$$= b\delta_{n0} + f_n(\theta)\,.$$

Finally, by reversing the integration order we get

$$\int_0^\theta \frac{\operatorname{tg}^{2n}\frac{\theta_1}{2}\, d\theta_1}{\sqrt{\cos\theta_1 - \cos\theta}} \int_{\theta_1}^\alpha \frac{g_n(\theta')\sin\theta'\, d\theta'}{\operatorname{tg}^n\frac{\theta'}{2}\sqrt{\cos\theta_1 - \cos\theta'}} =$$

$$= \frac{\operatorname{tg}^n\frac{\theta}{2}}{2a}\,(b\delta_{n0} + f_n(\theta))\,, \quad 0 \leqq \theta \leqq \alpha\,. \tag{7}$$

The solution of the integral equation (7) can be obtained by solving two successive ABEL type integral equations [3]

$$h_n(\theta) = \frac{1}{2a\pi\operatorname{tg}^{2n}\frac{\theta}{2}}\frac{d}{d\theta}\int_0^\theta \frac{\sin\theta'\operatorname{tg}^n\frac{\theta'}{2}\,(b\delta_{n0} + f_n(\theta'))\, d\theta'}{\sqrt{\cos\theta' - \cos\theta}} =$$

$$= \frac{\cos\frac{\theta}{2}}{\sqrt{2}a\pi}\, b\delta_{n0} + \frac{1}{2a\pi\operatorname{tg}^{2n}\frac{\theta}{2}}\frac{d}{d\theta}\int_0^\theta \frac{\sin\theta'\operatorname{tg}^n\frac{\theta'}{2} f_n(\theta')\, d\theta'}{\sqrt{\cos\theta' - \cos}}\,, \tag{8}$$

$$g_n(\theta) = -\frac{\operatorname{tg}^n\frac{\theta}{2}}{\pi\sin\theta}\frac{d}{d\theta}\int_\theta^\alpha \frac{\sin\theta'\, h_n(\theta')\, d\theta'}{\sqrt{\cos\theta - \cos\theta'}} \tag{9}$$

provided that the function $h_n(\theta)$ is continuously differentiable in $[\varepsilon, \alpha]$ for any positive ε. In the case $n = 0$, i.e. for an axialy symmetrical external field, the solution coincides with those given in [4]. The *total charge on the body* is

$$Q = 2\pi a^2 \int_0^\alpha g_0(\theta')\sin\theta'\, d\theta' =$$

$$= \frac{a(\alpha + \sin\alpha)}{\pi}\, b + 2\frac{\sqrt{2}a}{\pi}\int_0^\alpha \cos\frac{\theta}{2}\frac{d}{d\theta}\int_0^\theta \frac{\sin\theta' f_0(\theta')\, d\theta'}{\sqrt{\cos\theta' - \cos\theta}}\, d\theta'\,. \tag{10}$$

Relation (10) gives also the *body potential b* when the total charge Q is given.

The above presented solution is based on the integral relation (3) which has been proved by using the complex integration. Likewise the integral relation used in COMPSON's paper [1] can be simpler established.

References

1 COMPSON, E. T., On the problem of the electrified disc, Proc. Edinburgh Math. Soc., II. Ser 8 (1947), 14–19.
2 COLLINS, W. D., On the solution of some axisymmetric boundary value problems by means of integral equations, Quart. J. Mech. Appl. Math. 12 (1959), 232–241.
3 SNEDDON, I. N., Mixed Boundary Value Problems in Potential Theory, John Wiley, New York 1966.
4 KANWAL, R. P., Linear Integral Equations, Academic Press, New York 1971.

Eingereicht am 31. 10. 1979

Anschrift: Prof. Dr. D. HOMENTCOVSCHI, Central Institute of Mathematics, Academiei Street 14, R-Bucharest 1 — Romania

BUCHBESPRECHUNGEN

Göldner, H. (Hrsg.) u. a., Arbeitsbuch Höhere Festigkeitslehre. Elastizitätstheorie, Plastizitätstheorie, Viskoelastizitätstheorie. Leipzig, VEB Fachbuchverlag 1978. 367 S., 276 Abb., 10 Tab., 19,80 M, Auslandspreis 28,— M. Best.-Nr. 5462719

Höhere Festigkeitslehre, Elastizitätstheorie, Plastizitätstheorie und Viskoelastizitätstheorie zählen als Teilgebiete der höheren Mechanik an den Technischen Hochschulen und Fachhochschulen zu den obligatorischen Lehrgebieten in den Fachrichtungen des Maschineningenieurstudiums. Das vorliegende Buch bringt eine repräsentative Aufgabensammlung, die aus der Unterrichtspraxis an der Technischen Universität Dresden hervorgegangen ist. Die zur Lösung erforderlichen theoretischen Grundlagen werden allerdings nur in knapper Form (ohne Herleitungen) den einzelnen Abschnitten vorangestellt. Der Leser findet ca. 90 Aufgabentypen, die sich in das Schema der Strukturmodelle „Stab", „Scheibe", „Platte", „Schale" einordnen lassen und für Studenten wie auch für praktisch tätige Ingenieure als Anregungen dienen sollen. — Bei manchen Problemstellungen geht das Buch über sonst übliche Berechnungswege hinaus, und — je nach dem Schwierigkeitsgrad — werden auch geeignete Näherungsverfahren herangezogen (so z. B. bei der mittelstarken Platte).

Starnberg H. NEUBER

Burke, J. J. / Weiss, V. (Eds.), Nondestructive Evaluation of Materials. Sagamore Army Materials Research Conference Proceedings 23. New York-London, Plenum Press 1979. XI, 530 S., US $ 49.50. ISBN 0-306-40185-1

Der Band enthält 22 Übersichtsarbeiten (Kapitel) zur zerstörungsfreien Werkstoffprüfung, von denen allein 10 die röntgenographischen Methoden, ihre Geschichte und den gegenwärtigen Stand behandeln. L. V. AZÁROFF zieht in seinem einleitenden Aufsatz den Schluß, daß durch die Einführung intensitätsstärkerer Röntgenlichtquellen die Röntgenbeugung bei der Materialprüfung stärker als je zuvor eingesetzt wird. Die beschriebenen Anwendungsgebiete reichen vom Nachweis von Ermüdungsschäden (S. Taira und K. Kamachi) — H. K. Herglotz beschreibt ein historisches Beispiel hierzu in memoriam F. Regler, Wien, eines Pinoniers der zerstörungsfreien Werkstoffprüfung — über die Röntgentopographie zur Aufzeichnung von Gitterdefekten (S. Weissman), die Messung von Eigenspannungen (C. O. Ruud und G. D. Farmer) bis zur Röntgenblitzphotographie zur Untersuchung makroskopischen Materialverhaltens in kurzen Zeitintervallen und der neueren Röntgenblitzbeugung zur Untersuchung mikroskopischer Eigenschaften (Q. Johnson und D. Pellinen). Ähnlich wie Röntgenstrahlen werden Neutronen zur zerstörungsfreien Materialuntersuchung eingesetzt (J. John). Während Röntgenstrahlen bekanntlich von dichten Werkstoffen wie Metallen absorbiert werden, werden Neutronen von Werkstoffen absorbiert, die Wasserstoff enthalten. Das Verfahren eignet sich daher insbesondere zur Feststellung von Korrosion. L. S. Birks beschreibt den Stand der Röntgenspektroskopie zur quantitativen chemischen Analyse, die auf der Entdeckung der charakteristischen Röntgenspektren der chemischen Elemente durch Moseley im Jahre 1913 beruht. E. Macherauch gibt einen Überblick über den Stand der röntgenographischen Spannungsanalyse in Europa. Die Grundlagen der polychromatischen röntgenographischen Spannungsanalyse sowie einige Anwendungen bei Ermüdungsbruchproblemen werden von M. Nagao und S. Kusomoto behandelt. Sie führen für eine quantitative Versagensanalyse mit Röntgenstrahlen die Bezeichnung Röntgen-Fraktographie ein und zeigen, daß diese mehr Informationen als die photographische Fraktographie liefert.

Den zweiten Komplex stellen die Ultraschallmethoden mit vier Arbeiten dar. R. B. Thompson berichtet über die erzielten Ergebnisse bei der Bestimmung von Größe, Form und Orientierung von Defekten als Grundlage für eine Versagensvorhersage und (gemeinsam mit O. Buck) über die akustischen Wechselwirkungen mit inneren Spannungen in Metallen. Die beiden weiteren Arbeiten behandeln die elektromagnetischen Ultraschallgeber (T. J. Moran) und die Ultraschallspektroskopie (O. R. Gericke).

Einen dritten Komplex bilden die optischen Methoden. C. H. Palmer und R. E. Green geben eine vergleichende Darstellung der optischen Methoden zur Erfassung von Schallemissionssignalen und beschreiben ein akustisches Emissions-Interferometer. Als potentielles Verfahren der zerstörungsfreien Werkstoffprüfung — insbesondere zur Feststellung von Rißausbreitungen — wird von A. S. Kuo und H. W. Liu das Moiré-Verfahren beschrieben, ein in der Dehnungsanalyse bereits viel benutztes Verfahren. Neben der Bildverarbeitung im Rahmen der optischen Methoden werden noch einige weitere Methoden vorgestellt, so die magnetischen Methoden, die mit zu den ältesten in der zerstörungsfreien Werkstoffprüfung gehören und zur Charakterisierung von Rissen (z.B. Rißöffnungsverschiebung) in der Bruchmechanik und zur Lebensdauervorhersage beitragen können, jedoch theoretisch noch nicht genügend fundiert sind (J. R. Barton u. a.).

Das Buch gibt insgesamt einen guten Überblick über Stand und Leistungsfähigkeit der wichtigsten Methoden der zerstörungsfreien Werkstoffprüfung. Nahezu jedes Kapitel enthält außerdem ein umfangreiches Literaturverzeichnis.

Woltersdorf G. Haberland

S. Lukasiewicz, Local Loads in Plates and Shells. Monographs and Textbooks on Mechanics of Solids and Fluids. Mechanics of Surface Structures 4. Alphen aan den Rijn, Sijthoff & Noordhoff; Warszawa, PWN-Polish Scientific Publishers 1979. XVII, 569 S., 183 Abb., 18 Tab., Dfl. 170.—, US $ 85.—. ISBN 90-286-0047-7

Der Autor behandelt umfassend die Bestimmung von Spannungen in Platten und Schalen mit konzentrierten Belastungen. Das Buch beginnt mit zwei Kapiteln über die Grundgleichungen der Platten- und Schalentheorie, wobei vom Verhalten des dreidimensionalen elastischen Körpers ausgegangen wird und auch die in der klassischen Theorie vernachlässigten transversalen Schubdeformationen und Normalspannungen berücksichtigt werden. Platten und flache Schalen werden dabei gemeinsam abgehandelt, da die Plattengleichungen aus den Schalengleichungen hervorgehen, wenn die Krümmungsradien unendlich werden. Die allgemeinen Beziehungen für Platten, bei denen auch Normalkräfte in der Plattenebene berücksichtigt sind, gehen bei Vernachlässigung letzterer in die Reissnersche Theorie über. Aus dieser wiederum folgen die Beziehungen der klassischen Kirchhoffschen Theorie, wenn die transversalen Schubdeformationen vernachlässigt werden. Die Theorie der in der Technik häufig angewandten Sandwichplatten wird dabei in der klassischen linearen Form einbezogen. Die weiteren Kapitel behandeln die Lösungsmethoden für konzentrierte Belastungen, insbesondere die Anwendung von Fourierreihen und Fourierintegralen, wobei letztere zur Auffindung geschlossener Lösungen für Einzellasten im Innern dienen. Die Lösungsmethoden werden an verschiedenen Beispielen demonstriert, wie isotrope und orthotrope Kreisplatte, dicke Kreisplatte, Platte auf elastischer Unterlage, Belastung durch konzentriertes Biegemoment, thermische Singularitäten, Einflußfelder. Die konzentrierte Randbelastung wird am Beispiel der Kragplatte unter Berücksichtigung der Schubdeformationen gelöst. Nach der Darstellung der ebenen Elastizitätstheorie (Platten, die nur in ihrer Mittelebene belastet werden) werden Platten behandelt, die sowohl in der Ebene als auch senkrecht dazu auf Biegung beansprucht werden. Dazu gehören Platten mit großen Durchbiegungen, die sich wegen der großen Membranspannungen wie Schalen verhalten, und die Stabilitätsprobleme. Die Lösungen für Schalen beginnen mit der Membrantheorie, bei der Biegespannungen vernachlässigt werden. Es folgen Lösungen nach der allgemeinen Theorie für sphärische Schalen, bei denen Biegeeffekte mit der Entfernung vom Lastangriffspunkt schnell abklingen. Beliebige Schalen mit positiver Gaussscher Krümmung werden bei einer konzentrierten Belastung normal zur Mittelfläche mittels doppelter Fourierintegrale behandelt. Dazu beschreibt der Autor auch eigene experimentelle Untersuchungen. Ein Kapitel ist den Zylinderschalen gewidmet, ein weiteres den halbunendlichen Schalen mit konzentrierten Belastungen am freien Rand, wobei Durchbiegung und Spannungsfunktion wiederum durch Fourierintegrale dargestellt werden. Im vorletzten Kapitel werden große Durchbiegungen vornehmlich von sphärischen Schalen betrachtet. Das letzte und ziemlich umfangreiche Kapitel über den Entwurf von Platten und Schalen zeigt an verschiedenen Beispielen, wie hohe Spannungskonzentrationen vermieden werden, wenn bei Platten die Lasteintragung über geeignet geformte und angebrachte Balken erfolgt oder wie Optimierungsbedingungen bei Schalen behandelt werden. Das Buch ist vornehmlich für Ingenieure geschrieben, die dünnwandige Konstruktionen zu entwerfen haben. Komplizierte mathematische Darstellungen sind deshalb weitgehend vermieden. Trotzdem ist das Buch keineswegs nur ein Nachschlagewerk. Es weicht überdies, insbesondere was die einleitenden Kapitel über die Grundgleichungen anbetrifft, von vielen herkömmlichen Werken über Platten- und Schalentheorie ab. Wer z. B. verfeinerte Plattentheorien berücksichtigen will, dem ist dieses Buch zu empfehlen.

Woltersdorf G. Haberland

Müller, E. A. (Ed.), Mechanics of Sound Generation in Flows. Joint Symposium Göttingen, August 1979. IUTAM/ICA/AIAA-Symposium Göttingen 1979. Berlin-Heidelberg-New York, Springer-Verlag 1979. XV, 300 S., 177 Abb., DM 68,—. US $ 38.10. ISBN 3-540-09785-6

Der vorliegende Sammelband enthält den Wortlaut der 38 Vorträge, die auf dem gemeinsam von der IUTAM, der ICA und dem AIAA veranstalteten Symposium über die Schallerzeugung in Strömungen gehalten wurden. Der weitaus größte Teil der mitgeteilten Ergebnisse behandelt die Schallentstehung im Freistrahl und damit im Zusammenhang stehende Fragen, wie die Wechselwirkung zwischen angeregten Störungen, festen, in der Nähe befindlichen Wänden, Wirbelkonfigurationen u. a. mit dem turbulenten Freistrahl (17 Arbeiten). Die übrigen Vorträge verteilen sich auf Fragen der Schallerzeugung an der Hinterkante von Tragflügeln, durch Propeller, Fragen nach der Bedeutung verschiedener Wirbelformationen für die Schallentstehung u. a. m. Die fünf Einladungsvorträge wurden von D. G. Crighton zu Problemen der Berechnung von großskaligen Strukturen und der Akustik der mittleren Strömung, von W. Möhring zur Modellierung des Lärms bei kleiner Machzahl, von F.-G. Grosche zum gegenwärtigen Stand der Experimentaltechnik, von R. J. Hill zu den Erwartungen der Industrie gegenüber der Grundlagenforschung und von D. B. Hanson zur Aeroakustik von modernen Turbopropellern gehalten. Auf den Inhalt der einzelnen Beiträge kann hier nicht detailliert eingegangen werden;

dem am Gegenstand Interessierten wird deshalb die anregende Lektüre des Materials empfohlen, die wegen des dokumentierten aktuellen Standes der Forschung auf dem Gebiet der Strömungsakustik den Fachkollegen in aller Welt gestattet, ihre eigene Position zu überprüfen. Ebenso erstaunlich wie die schnelle Herausgabe ist auch die für eine derart aktuelle Information ungewöhnliche gute Ausstattung des Bandes.

Berlin K. Remke

Pedlosky, J., Geophysical Fluid Dynamics. New York-Heidelberg-Berlin, Springer-Verlag 1979. XII, 624 S., 180 Abb., DM 79,50. US $ 43.80. ISBN 3-540-90368-2

Dies ist ein Lehrbuch der Strömungslehre für die Atmosphäre und den Ozean. Die hydrodynamischen Grundgleichungen werden von Anfang an unter dem Gesichtspunkt betrachtet, daß es sich um dünne Flüssigkeitsschichten auf einer rotierenden Kugel handelt und sie sich durch Größenordnungsabschätzungen nach Bedarf vereinfachen lassen. Der Verfasser setzt sich das Ziel, diese vereinfachten Gleichungen einer sorgfältigen und genauen Analyse zu unterziehen. Er sagt, die Vereinfachungen sind notwendig, um einen Fortschritt möglich zu machen, während die Genauigkeit nötig ist, damit der Fortschritt sinnvoll wird. Das Buch ist klar, manchmal sogar eindringlich geschrieben und schon für Studenten höherer Semester lesbar. Im Vordergrund stehen die mathematisch-physikalischen Grundlagen, Erfahrungstatsachen dienen lediglich als Motivation. Es handelt sich also um kein Lehrbuch der Ozeanologie oder der Meteorologie, und auch der numerische Aspekt steht außerhalb der Betrachtung. Wellenentstehung und Wellenausbreitung bilden den Schwerpunkt. Die zentralen Kapitel behandeln die Flachwassertheorie einer idealen Flüssigkeit, den Einfluß von Zähigkeit und Reibung, homogene Modelle für windgetriebene Meereszirkulation, quasigeostrophische Bewegung einer geschichteten Flüssigkeit auf der Kugel und die Instabilitätstheorie.

Wer um ein tieferes Verständnis von Luft- und Meeresströmungen bemüht ist, wird das vorliegende Buch mit Gewinn studieren.

Berlin J. Förste

Fachkolloquium 1979, Möglichkeiten mathematischer Modelle bei der Lösung von Strömungsproblemen in Umweltschutz und Wasserbau, veranstaltet von den Sonderforschungsbereichen SFB 79, Universität Hannover; SFB 80, Universität Karlsruhe; SFB 81, TU München; SFB 94 Universität Hamburg, Karlsruhe, den 24.–25. 9. 1979. SFB 80, Universität Karlsruhe, Kaiserstraße 12, D-7500 Karlsruhe. 191 S.

An Hand aktueller hydrologischer Probleme in der BRD werden in diesem Tagungsbericht Möglichkeiten für die mathematische Behandlung von Strömungsproblemen erörtert. Dabei ergibt sich die inhaltliche Gliederung nach den Programmen der dort tätigen, interdisziplinär angelegten, sogenannten Sonderforschungsbereiche (SFB), in denen Wissenschaftler aus verschiedenen Institutionen zusammengefaßt sind. Es handelt sich um die 4 SFB: Wasserforschung im Küstenbereich, Ausbreitung und Transportvorgänge in Strömungen, qualitative und quantitative Abflußvorgänge in natürlichen und künstlichen Gerinnen mit besonderer Berücksichtigung der Verhältnisse im alpinen und voralpinen Raum und Meeresforschung (Impuls-, Energie- und Stofftransport im System Ozean/Atmosphäre).

Das Heft enthält insbesondere Beiträge über Möglichkeiten mathematischer Modelle bei der Lösung von Strömungsproblemen in Tidegewässern und im Meer, in Flüssen, Seen und der Atmosphäre und im Grundwasser. Bei der Auswahl der Berechnungsverfahren stehen neben wissenschaftlichen vor allem ökonomische Gesichtspunkte im Vordergrund.

Berlin J. Förste

Fife, P. C., Mathematical Aspects of Reacting and Diffusing Systems. Lecture Notes in Biomathematics 28. Berlin-Heidelberg-New York, Springer-Verlag 1979. IV, 185 S., 23 Abb., DM 21,50. US $ 11.90. ISBN 3-540-09117-3

Die Untersuchung der Dynamik chemischer Mischungen durch Reaktions-Diffusions-Gleichungen (nichtlineare parabolische Differentialgleichungen) stellt eines der wohldefinierten und am Experiment direkt überprüfbaren Probleme der Biomathematik dar. Ausgerüstet mit den Ergebnissen einer Reihe gut ausgebauter Methoden (Vergleichs- und Stabilitäts-, Bifurkations-, singuläre Störungs- und Skalierungstechniken) können die bei skalaren Gleichungen erfolgreichen Verfahren auf die Untersuchung mehrkomponentiger Größen und die entsprechenden Gleichungssysteme erweitert werden. Ein interessantes Beispiel der untersuchten dynamischen Erscheinungen stellen die bekannten dissipativen Strukturen dar.

In die klare und übersichtliche Monographie von P. C. Fife fließen auch wesentliche Ergebnisse des Autors auf dem Gebiet der Musterbildung in Reaktions-Diffusions-Systemen sowie der Bistabilität und Asymptotik in solchen Systemen ein. Die sehr wertvolle Literatur (über 300 Zitate) ist gewissenhaft verarbeitet, so daß dem Text weit über den eigentlichen Umfang der einzelnen Kapitel hinausgehende Hinweise zu entnehmen sind.

Berlin G. Wallis

Einstein Symposion Berlin, 25. bis 30. 3. 1979. Eds.: Nelkowski, H./Hermann, A./Poser, H./Schrader, R./Seiler, R., Lecture Notes in Physics 100. Berlin-Heidelberg-New York, Springer-Verlag 1979. VIII, 550 S., 35 Abb., DM 49,50. US $ 27.80. ISBN 3-540-09718-X

Der umfangreiche Band enthält die Ansprachen und Vorträge, die auf dem Einstein-Symposium in Berlin (West) aus Anlaß des 100. Geburtstages von Albert Einstein gehalten wurden. Träger dieser Veranstaltung waren der Senator für Wissenschaft und Forschung, die FU sowie die TU. Die Beiträge zur Gravitationstheorie wurden von J. Ehlers (München) und R. Penrose (Oxford) geliefert. Zur Einheitlichen Feldtheorie sprachen P. G. Bergmann (Syracuse), J. Iliopoulos (Paris) und B. Zumino (Genf). Des weiteren waren mehrere Vorträge den Mikroprozessen und stochastischen Prozessen gewidmet. Diese drei Themenkomplexe, die auch in der Forschung Einsteins im Vordergrund standen, wurden durch vier Workshops ergänzt: Feldtheorie (Vor- und Rückschau), Kosmologie, Struktur und Entwicklung physikalischer Theorien, Einstein — Mensch und sozialer Kontext. Die Gesamtheit der Vorträge korrespondiert dem gesetzten Ziel, Wert und Wirken Einsteins in physikalischer, wissenschaftstheoretischer und historischer Sicht zu beleuchten. Da führende Vertreter der internationalen Wissenschaft als Redner gewonnen wurden, ist dieses Ziel auch weitgehend erreicht worden. Allerdings dürften die fachspezifischen Beiträge den allgemein-wissenschaftlichen Rahmen des Symposiums ziemlich gesprengt haben, da sie zu sehr ins Detail gehen. Für den Fachmann selbst sind diese Beiträge natürlich von besonderem Nutzen, weil er den neusten Stand der Forschung von kompetenten Vertretern erfährt. Insgesamt gesehen wird ein wertvolles Werk zur Einstein-Literatur angeboten, das besonders durch seine Vielseitigkeit beeindruckt.

Jena E. Schmutzer

Thirring, W., Lehrbuch der Mathematischen Physik, Band 3: Quantenmechanik von Atomen und Molekülen. Wien-New York, Springer-Verlag 1979. X, 263 S., 22 Abb., S 250,–; DM 36,–; US $ 19.80. ISBN 3-211-81538-4

Der vorliegende Band ist der dritte Teil eines auf vier Bände angelegten Lehrbuches der Mathematischen Physik (siehe die entsprechenden Referate der vorhergehenden Teile).

Auch in diesem, der Quantenmechanik von Atomen und Molekülen gewidmeten Band versteht es der Autor in überzeugender Weise, eine moderne mathematische Darstellung (sowohl der Grundlagen der Theorie als auch von konkreten Einzelsystemen) mit physikalisch relevanten Erörterungen zu verbinden. Die Darstellung führt an ziemlich aktuelle Fragestellungen heran und verarbeitet neueste, beim Erscheinen des Buches z. T. erst in Preprint-Form vorliegende Literatur.

Das Leitmotiv der Darstellung ist auch in diesem Band das erfolgreiche Bestreben, nur mathematisch exakte Ableitungen und Methoden zu bringen, die auf einer gründlichen mathematischen Analyse der Grundstruktur (hier der Schrödinger-Gleichung) beruhen, aber nach Möglichkeit bis zu numerischen Resultaten vorzudringen. Demzufolge nehmen exakte Approximationsmethoden einen breiten Raum ein.

Nach einer heuristischen Einleitung, in der bereits darauf hingewiesen wird, daß es z. B. bei der Bewältigung der Stabilitätsprobleme der Quantenmechanik von Atomen und Molekülen auf eine „feingeschliffene Rechentechnik" ankommt, wird in Kap. 2 der mathematische Apparat der Quantenmechanik skizziert (C^*- und W^*-Algebren, (normale) Zustände, GNS-Darstellung, Spur, Operator(halb)gruppen, unbeschränkte Operatoren, Formen, Erweiterungstheorie). Im Kap. 3 folgen dann nach Einführung WEYLscher Systeme und der Behandlung des Drehimpulses die Theorie der quantenmechanischen Zeitevolution, die Grundlagen der zeitabhängigen ($t \to \pm\infty$) und der stationären Streutheorie sowie die Störungstheorie diskreter Spektren. Kap. 4 enthält die Anwendung der Theorie auf verschiedene konkrete Systeme (Wasserstoffatom, Wasserstoffatom in äußeren Feldern, Heliumatom u. a.). Mathematisch läßt sich dieses Kapitel als qualitative und quantitative Spektraltheorie der entsprechenden SCHRÖDINGER-Operatoren einordnen. Dabei wird die von der Spektraltheorie kommende Begrifflichkeit und Systematik sehr ernst genommen. Z. B. werden Abwesenheitsbeweise für singulärstetiges Spektrum gebracht, häufig unter Benutzung der sog. dilatationsanalytischen Fortsetzungstechniken. Bei der Behandlung des Starkeffektes wird deutlich herausgestellt, daß die angewendete Störungstheorie in diesem Fall eigentlich gar nicht anwendbar ist, weil die in Rede stehenden Eigenwerte instabil sind, die „formalen" Störungsreihen jedoch, u. a. mit Hilfe des Begriffes der Spektralkonzentration, real interpretiert werden können.

Alles in allem stellt das Buch, in dem nach Kenntnis des Referenten viele neuere Methoden zur Behandlung der SCHRÖDINGER-Gleichung zum ersten Male zusammenfassend lehrbuchmäßig dargestellt sind, eine echte Bereicherung der wahrlich nicht gerade spärlichen Lehrbuchliteratur zur Quantenmechanik dar.

Berlin H. BAUMGÄRTEL

Wentzell, A. D., Theorie zufälliger Prozesse. Mathematische Monographien 50. Berlin, Akademie-Verlag 1979. X, 253 S., M 48,—. BN 7624531 (6434)

Dieses Buch ist ein aus Vorlesungen des Autors hervorgegangenes Lehrbuch; es enthält sehr viel Material, das bislang nur in Form von Monographien veröffentlicht worden ist. Nach einer Einführung in die grundlegenden Begriffe, in der eine grobe Übersicht über die Methoden der Theorie zufälliger Prozesse gegeben wird, wird die Korrelationstheorie und Spektraltheorie entwickelt. So vorbereitet wird der Leser an den Fundamentalsatz von KOLMOGOROV herangeführt. Die allgemeine Theorie zufälliger Prozesse wird dann mit Kapiteln über MARKOV-Zeiten und Martingale abgeschlossen. Die verbleibende Hälfte des Buches behandelt die (moderne) Theorie der MARKOV-Prozesse in großer Ausführlichkeit. Insbesondere findet der Zusammenhang von MARKOV-Prozessen mit der Halbgruppentheorie von HILLE-YOSIDA eine in sich abgeschlossene Darstellung. Daran schließt sich an eine Einführung in die Theorie der Diffusionsprozesse und eine Einführung in die Theorie der stochastischen (ITÔ-) Differentialgleichungen. Das Buch endet mit dem Zusammenhang zwischen Diffusionsprozessen und partiellen Differentialgleichungen.

Die Fülle von (anwendungsorientierten) Beispielen und Aufgaben, deren Lösungen man vollständig im Anhang findet, machen dieses sehr verständlich geschriebene Lehrbuch in hervorragender Weise zum Selbststudium geeignet. An Vorkenntnissen werden Maßtheorie, (lineare) Funktionalanalysis und elementare Wahrscheinlichkeitstheorie vorausgesetzt. Das Buch richtet sich daher vorwiegend an Mathematiker, Mathematik-Studenten höheren Semesters und an Anwender mit entsprechender Vorbildung.

Bremen L. ARNOLD

Kijowski, J. / Tulczyjew, W. M., A Symplectic Framework for Field Theories. Lecture Notes in Physics 107. Berlin-Heidelberg-New York, Springer-Verlag 1979. IV, 257 S., DM 28,50. US $ 15.70. ISBN 3-540-09538-1

Die Untersuchung symplektischer Strukturen hat in den letzten Jahren in der theoretischen Physik zunehmend Interesse gefunden, insbesondere im Rahmen der klassischen Punktmechanik, der Quantentheorie und der Zusammenhänge beider. Unter einer symplektischen Mannigfaltigkeit versteht man dabei etwa eine differenzierbare Mannigfaltigkeit zusammen mit einer Cartanschen Differentialform oder einer anderen symplektischen Form. Ein besonders wichtiger symplektischer Raum ist der Phasenraum der klassischen Mechanik und Statistik; er besitzt infolge der Gültigkeit der Hamiltonschen kanonischen Differentialgleichungen eine symplektische Struktur und bildet auch den Ausgangspunkt der Entwicklungen des vorliegenden Buches.

Dieses Buch trägt teils Übersichtscharakter, teils ist es eine Originalarbeit. Sein Anliegen ist eine neuartige Darstellung klassischer Feldtheorien — Quantenfelder werden nicht betrachtet — und insbesondere die Entwicklung neuer physikalischer Ideen, wobei die kononische Formulierung von Feldtheorien, die Beziehungen von LAGRANGE- und HAMILTON-Systemen sowie Variationsprinzipien im Vordergrund stehen. So wird ein systematisches Verfahren zur Herleitung von Variationsprinzipien angegeben, und neue Variationsprinzipien werden für die Hydrodynamik und die Gravitationstheorie aufgestellt. Dissipative Prozesse bleiben außerhalb der Betrachtungen.

Von der abstrakten Sprache der modernen Differentialgeometrie wird nur mäßiger Gebrauch gemacht. Es erwies sich als zweckmäßig, den Begriff der symplektischen Struktur für Felder etwas allgemeiner zu fassen. Die damit zusammenhängenden mathematischen und physikalischen Probleme finden eine ausführliche Darstellung. Dabei gehen die Autoren sehr systematisch vor und bringen nach einer intuitiven Einführung und Übersicht über die symplektische Geometrie in der Physik zunächst die bekannten Probleme der nichtrelativistischen Punktmechanik in neuer Sicht, um danach ausgiebig Fragen der nichtrelativistischen und relativistischen Feldtheorien zu diskutieren. Als Anwendungsbeispiele werden gewöhnliche Vektorfelder und PROCA-Felder, das elektromagnetische Feld, das Gravitationsfeld und die Hydrodynamik untersucht.

Das Buch kann ohne einige Vorkenntnisse über Differentialgeometrie und Differentialformen nicht gelesen werden. Es ist ein sehr bedeutsamer Beitrag zu einem aktuellen Forschungsgebiet der modernen mathematischen Physik.

Dresden G. VOJTA

Jacod, J., Calcul Stochastique et Problèmes de Martingales. Lecture Notes in Mathematics 714. Berlin-Heidelberg-New York, Springer-Verlag 1979. X, 539 S., DM 46,—. US $ 25.30. ISBN 3-540-09253-6

This monography is an excellent systematic presentation of the theory of martingales and stochastic integrals. It is the first book on stochastic analysis which contains more or less all fundamental results of this theory up to 1979. On the basis of the general theory of processes developed in the Strasbourg school the author presents in a systematic, mathematically exact and complete form the theory. Because of his encyclopaedic treatment of the subject the book could become one of the leading reference works in stochastic analysis. A disadvantage is that it is written in French. The monography is selfcontaining. During first reading one gets the impression that the presented topics are almost completely solved. But this is generally not true and especially not concerning the last chapters. For a student the use of the book might be difficult, because only a few remarks are given which explain the background of the theory and the results. But after an introductory course in stochastic analysis the book will be very useful also for students and other scientists working on that field. At the end of each chapter the author formulates exercises which deepen the understanding and develop the theory on special questions.

The first part of the book presents the fundamental theory. It is a summary of the general theory of processes developed by the Strasbourg school. Then stochastic integration with respect to martingales and semimartingales is developed. Probably, here is also given the first systematic presentation of the theory of random measures in stochastic analysis with application in stochastic integration. In the second part the author studies transformations of processes connected with the exponential formula, absolute continuity and change of filtration, time or space. The third part treats martingale problems and stochastic differential equations. At the end of the book an almost complete bibliography of the subject is given. It enables the reader to find more background material and detailed discussions about special problems he might be interested in. In fact it was impossible to include also such material in the monography which reached already more than 500 pages.

The book can be recommended to mathematicians, students and other scientists working in the field of stochastic analysis and having some knowledge on measure theory and a little experience of martingales and stochastic integration.

Berlin E. PLATEN

Barut, A. O. (Ed), Nonlinear Equations in Physics and Mathematics. Proceedings of the NATO Advanced Study Institute held in Istanbul, Turkey, August 1—13, 1977. NATO Advanced Study Inctitute Series. Series C— Mathematical and Physical Sciences 40. Dordrecht-Boston-London, D. Reidel Publ Co. 1977. X, 473 S., 70 Abb., Dfl. 80,—. US $ 39.00. ISBN 90-277-9936-X

Eine Abgrenzung gegenüber allgemeinen nichtlinearen Gleichungen in Physik und Mathematik, wie sie der Titel dieses Tagungsbandes vielleicht erwarten läßt, liegt in der Beschränkung auf die vollständig integrierbaren Systeme, welche der Behandlung durch die inverse Streu-Transformation zugänglich sind (Teil I). Interessante Gegenüberstellungen liefern die den klassischen Systemen gewidmeten Beiträge und diejenigen zu quantisierten Feldern. Im ersten Fall (Teil III) sind es Vielkörperprobleme, Fragen der nichtlinearen Gitterdynamik und der nichtlokalen Kontinuumsmechanik (zweidimensionale Ising-Modell, Phasenübergänge, Versetzungen und nichtlokale Elastizitätsmoduln). Teil IV stellt sich die Aufgabe, Quantisierungen bzw. „charakteristische Quanten" nichtlinearer Feldgleichungen zu finden (Thirring-Modell, nichtlineare duale Felder, kollektive Phänomene). Da es sich um nicht-störungstheoretische Verfahren handelt, sind auch neue Lösungstypen zu erwarten. Diese liegen in der Form der „Solitonen" vor, denen ebenfalls ein Abschnitt (Teil II) gewidmet ist (differentialgeometrische Interpretation, Hirotas direkte Methode, Anwendungen).

Spezielle mathematische Methoden sind in diesem Zusammenhang die Bäcklund- und die Hopf-Cole-Transformation, spezielle Gleichungen, die Gelfand-Levitan-Marchenko-Gleichung, die Lax-Gleichung, die Painlevé-Gleichung dritter Art, das verallgemeinerte Zakharov-Shabat-Problem.

So steht die überdeckte Thematik in einer Mittelstellung zwischen mathematisch-methodischen Untersuchungen nichtlinearer partieller Differentialgleichungen und recht detaillierten Anwendungen auf verschiedene physikalische Systeme (Festkörper, Plasmen, Supraleiter und Systeme der nichtlinearen Optik).

Die Vielfalt der angesprochenen physikalischen Systeme, der hohe Schwierigkeitsgrad und die Spezifität der bei jedem dieser Systeme zu findenden nichtlinearen Phänomene weisen auch auf die Problematik einer Veranstaltung hin, nichtlineare Gleichungen in Physik und Mathematik unter einem gemeinsamen Gesichtspunkt zu behandeln.

Berlin G. Wallis

Rao, M. M., Stochastic Processes and Integration. Alphen aan den Rijn, Sijthoff & Noordhoff 1979. XI, 456 S., Dfl. 110.00, US $ 55.00. ISBN 90-286-0438-3

This monography is presenting foundations and central parts of stochastic analysis. It is written on a high theoretical level. The author starts in most chapters with more easily accessible material and progresses later to more abstraction. So the motivation becomes clear which makes the book useful for graduate students as well as for research workers. Obviously, the author prefers functional analytic methods to develop the theory which shows also the traditional influence of functional analysis on stochastic analysis. Many of his own results are included in the text.

The book starts with the fundamental existence theorem of Kolmogorov in an abstract formulation. Some of its extensions as projective limits and a few applications follow. Then conditional expectations are studied more detailed as usual. In the following he presents martingales, their decomposition, martingale inequalities and convergence theorems. The concepts of separability and measurability are introduced and analysed for general classes of processes. Here, the author is using the so-called lifting theory, where lifting means a continuous selection from equivalence classes in a topological space. Together with stopping time transformations various properties of martingales are studied. Furthermore, several decompositions of processes are derived and the space of bounded mean oscillation is studied. It follows the important Doob-Meyer decomposition, which allows the consideration of square integrable martingales and the introduction of the Itô integral. In a special section the relation between stochastic and spectral integrals is pointed out. In the chapter the concept of semimartingales is developed. Finally a sketch of multiparameter martingales is added together with indications of current trends in stochastic analysis.

At the end of each chapter complements and problems are given which contain additional results. The reader must be familiar with standard real analysis and Hilbert space theory. Where special knowledge is needed the author is giving references. An extensive bibliography concludes the book.

Berlin E. Platen

Braun, M., Differentialgleichungen und ihre Anwendungen. Übers. a. d. Engl. von T. Tremmel. Hochschultext. Berlin-Heidelberg-New York, Springer-Verlag 1979. XII, 596 S., 65 Abb., 17 Tab., DM 48,—, US $ 26.40. ISBN 3-540-09391-5

Dieses originelle Lehrbuch hat einführenden Charakter; es wendet sich auch an Studenten der Anwendungsgebiete; es erreicht sein Ziel mit einem Minimum an mathematischem Aufwand von einem Minimum an Vorkenntnissen aus. Wo der Autor da Lücken beim Leser vermutet, fügt er an sich nicht zum Thema gehörende Exkurse ein. — Die Kapitelüberschriften: 1. Differentialgleichungen erster Ordnung, 2. Lineare Differentialgleichungen zweiter Ordnung, 3. Systeme von Differentialgleichungen, 4. Qualitative Theorie der Differentialgleichungen, 5. Separation der Variablen und Fourierreihen. Das fünfte Kapitel bringt hauptsächlich die Methode der Variablentrennung zur Lösung der einfachsten Randwert- bzw. Anfangs-Randwertaufgaben für die drei klassischen Grundtypen partieller Differentialgleichungen, also der Laplace-, der Wärmeleitungs-, der Wellengleichung. Die anderen Kapitel behandeln gewöhnliche Differentialgleichungen nach bekannten Mustern: Hauptthemen in Kap. 1 sind lineare Gleichungen, Gleichungen mit trennbaren Veränderlichen, exakte Gleichungen, Existenz- und Eindeutigkeitsaussagen, numerische Approximationen, z. B. die Verfahren von Euler und Runge-Kutta. Kap. 2 behandelt hauptsächlich Potenzreihenansätze, Gleichungen mit konstanten Koeffizienten und die Laplacetransformation. (Außer linearen sollten vielleicht doch die elementar behandelbaren Differentialgleichungen 2. Ordnung auch in einer Einführung nicht fehlen.) In Kap. 3 werden lineare Systeme mit konstanten Koeffizienten betrachtet und dafür Elementarstoff aus der linearen Algebra bereitgestellt. Kap. 4 ist Stabilitätsfragen gewidmet. Das Buch enthält an die tausend Übungsaufgaben, die Hälfte mit Lösungen. Zu erwähnen ist ferner ein Anhang, der in die Programmiersprache APL einführt, wie überhaupt die Darstellung dem Problem der numerischen Realisierung von Ergebnissen gebührende Beachtung schenkt. — Das Buch ist insgesamt solide aufgebaut, im Detail aber hätte, ohne die Gesamtanlage zu ändern, vieles präziser formuliert werden sollen. Das beginnt schon bei den Kapitelüberschriften. Einige wenige weitere Beispiele für die ungenügende Selbstredaktion durch den Autor: Anders als im Titel heißt es im ersten Satz des Vorwortes, das Buch behandele die Theorie der gewöhnlichen Differentialgleichungen; der Anfänger, dem der Terminus „gewöhnlich" nichts sagt, wird durch das Sachverzeichnis auf die Definition S. 1 verwiesen, findet dort indessen wie auch auf den folgenden Seiten nichts; erst im Laufe des 5. Kap., auf S. 518 wird unterstellt, der Leser kenne den hier erstmals auftauchenden Begriff. Manchmal werden Iterationen a priori als Approximationen angesprochen (z. B. S. 93 oben). Die Begriffe „allgemeine" bzw. „partikuläre" bzw. „singuläre" Lösung werden nur an Beispielen oder gar nicht erläutert, es gibt keine Definition. In der Formel auf S. 109 unten wird plötzlich die Größe h benutzt, aber erst später im Text eingeführt. Wichtige Sätze sind oft durch Kursivsatz hervorgehoben, manchmal aber muß man sie suchen (Bsp. S. 116, Formel (6)). Oder man sehe Formulierungen wie S. 2 unten/S. 3 oben: „Um eine beliebige Differentialgleichung lösen zu können, müssen wir sie auf irgendeine Weise auf die Form $dy/dt = g(t)$ bringen. In den meisten Fällen ist dies aber unmöglich, wie in Abschnitt 1.9 gezeigt wird. Die meisten Gleichungen lassen sich daher nur mit Hilfe eines Computers lösen." Hier kann auch der talentierteste Student unmöglich erraten (es sei denn, er weiß es schon aus anderen Quellen), was der Autor meint, höchstens, daß letzterer von Computern träumt, die Unmögliches doch noch möglich machen. So darf zu den eingangs genannten, dem Anfänger jedenfalls dienlichen Minima kein weiteres an Intuition hinzukommen. Oder, um die Kritik ins Positive zu wenden: Der Leser wird durchaus nicht im Bourbakismus erstickt. — Alle Mängel des Buches werden jedoch mehr als wettgemacht durch seinen größten Vorzug: Eine Fülle von weit ausgeführten, interessant vorgetragenen, jedermann ansprechenden Beispielen, die nicht als obligates Anhängsel, sondern als integrierender Bestandteil der Darstellung wirken und zweierlei zeigen: 1. Die Sache ist nützlich, 2. die Sache macht Spaß. Um was geht es da nicht alles: Um Kunstfälschungen, Atommüllbeseitigung und Brückeneinstürze, um den Kampf um die Insel Iwoshima 1945 und den Haifischfang im Mittelmeer, um die Ausbreitung von Geschlechtskrankheiten und um einen Pseudovektorraum aus Hunden, Katzen und Mäusen, um nur einiges zu nennen. Man merkt, daß es auch dem Autor Spaß gemacht hat, dieses Buch zu schreiben, sonst hätte er kaum drei Seiten vor Schluß einen Computer als Beispiel den Satz „Martin Braun is a great guy" ausdrucken lassen. Es lohnt sich, das Buch in die Hand zu nehmen und festzustellen, ob dieser Satz richtig ist.

Berlin W. HEINRICH

Gurel, O. / Rössler, E., Bifurcation Theory and Applications in Scientific Disciplines. Annals 316. New York, The New York Academy of Sciences 1979. 708 S., US $ 85.00. ISBN 0-89766-558

Die Bifurkationstheorie befaßt sich mit den qualitativen Veränderungen im Lösungsverhalten nichtlinearer Systeme bei Einwirkung von Störungen, die im allgemeinen durch Änderungen gewisser Systemparameter realisiert werden. Ein entscheidender Schritt zur Herausbildung dieser Theorie wurde 1942 von E. HOPF getan, der die Abzweigung einer periodischen Lösung aus dem Gleichgewichtspunkt eines n-dimensionalen autonomen Systems parameterabhängiger gewöhnlicher Differentialgleichungen untersuchte. Diese unter dem Namen HOPF-Bifurkation bekannte Erscheinung konnte später auch bei partiellen Differentialgleichungen nachgewiesen und in mannigfacher Weise verallgemeinert werden.

1977 wurde aus Anlaß des 75. Geburtstages von E. HOPF von der New Yorker Akademie der Wissenschaften die bisher bedeutendste und universalste Konferenz über Bifurkationstheorie und deren Anwendungen organisiert. An dieser Tagung nahmen nahezu alle führenden Vertreter der Bifurkationstheorie aus der westlichen Hemisphäre teil (u. a. auch E. LORENZ und P. H. RABINOWITZ, deren Namen im Autorenverzeichnis nicht auftauchen), als prominenter Vertreter dieser Theorie fehlte wohl nur R. THOM, der aber um so häufiger zitiert wurde.

Die Anwendungsbreite der Bifurkationstheorie ist enorm. Bemerkenswert ist, daß in jüngster Zeit die Untersuchungen über turbulente Erscheinungen im Rahmen dieser Theorie an Bedeutung und Umfang zugenommen haben.

Der vorliegende Band enthält sowohl Beiträge zu rein mathematischen Fragestellungen der Bifurkationstheorie als auch zu Anwendungen derselben in Biologie, Biochemie, Chemie, Hydrodynamik, Ökonomie, Ökologie, Technik. Von den Autoren seien u. a. die Namen R. H. ABRAHAM, J. F. AUCHMUTY, M. CONRAD, J. CRONIN, H. DEGN, P. R. FENSTERMACHER, M. GOLUBITZKY, J. GUCKENHEIMER, O. GUREL, H. HAKEN, J. K. HALE, B. HESS, F. C. HOPPENSTEADT, PH. HOLMES, G. W. HUNT, D. D. JOSEPH, R. LANDAUER, M. C. MACKEY, R. M. MAY, J. E. MARSDEN, S. E. NEWHOUSE, G. NICOLIS, Z. NITECKI, P. ORTOLEVA, H. G. OTHMER, O. E. RÖSSLER, R. ROSEN, D. H. SATTINGER, F. F. SEELIG, ST. SMALE, J. M. T. THOMPSON, J. J. TYSON, R. F. WILLIAMS erwähnt.

Berlin K. R. SCHNEIDER

Löhr, K. J., Beispiele und Aufgaben zur Laplace-Transformation mit 167 Beispielen und 37 Aufgaben. Viewegs Fachbücher der Technik. Braunschweig-Wiesbaden, Friedr. Vieweg & Sohn 1979. VI, 202 S., 61 Abb., 1 Tab., DM 24,80. ISBN 3-528-04082-3

Das Buch hat das Ziel, Studenten Technischer Hochschulen und Fachhochschulen schon im zweiten Studienjahr mit der Theorie und Praxis der LAPLACEtransformation bekannt zu machen, so daß sie gewöhnliche Differentialgleichungen mit konstanten Koeffizienten lösen können und eine Grundlage für das Studium weiterführender Werke über die Laplacetransformation bei fortschreitender Kenntnis in der Mathematik erhalten. Nach der ausführlichen Berechnung zahlreicher LAPLACE-transformierter werden einige allgemeine Eigenschaften der LAPLACEtransformation wie Verschiebungssätze und Faltungssatz eingeführt und wiederum anhand instruktiver Beispiele erläutert, wobei auch ausführlich auf Möglichkeiten fehlerhaften Vorgehens hingewiesen wird. Bei der Lösung gewöhnlicher Differentialgleichungen mit Hilfe der LAPLACEtransformation wird wiederholend auf die üblichen Lösungsmethoden, ebenso auf die Partialbruchzerlegung für die Rücktransformation eingegangen. Die gewonnenen Lösungen von Differentialgleichungen zweiter Ordnung mit konstanten Koeffizienten werden als Schwingungsmöglichkeiten ausführlich interpretiert. An einigen Beispielen wird gezeigt, daß auch Integrodifferentialgleichungen in derselben Weise gelöst werden können. Das Buch schließt mit Anwendungen auf Stromkreise, auf die Durchbiegung von Balken und, als Beispiele für Systeme gewöhnlicher Differentialgleichungen, auf elektrische Vierpole. Zahlreiche Übungsaufgaben mit Lösungen runden das Buch ab, das neben dem angesprochenen Leserkreis wohl auch schon interessierten Lesern mit mathematischen Abiturkenntnissen empfohlen werden kann.

Berlin G. SCHMIDT

Franksen, O. I. / Falster, P. / Evans, F. J., Qualitative Aspects of Large Scale Systems. Developing Design Rules Using APL. Lectures Notes in Control and Information Sciences 17. Berlin-Heidelberg-New York, Springer-Verlag 1979. XII, 119 S., DM 18,–, US $ 9.90. ISBN 3-540-09609-4

The increase in size and complexity combined with the need for taking non-linearity into account, call for new concepts and procedures in the design process of automatic control systems. To meet this need for a maximum rationalisation of behaviour in the design process research must be undertaken to develop qualitative or structural design rules which, in the form of computer-aided methods, may be applied in the design of high-order and non-linear automatic control systems without actually attempting explicit solutions.

The author of this monograph is concerned with a set of design concepts that represents, in a qualitative manner, the properties or controllability and observability of timeinvariant systems.

Berlin G. TELCHOW

Siegert, H. J. (Hrsg.), Virtuelle Maschinen. Nachbildung und Vervielfachung maschinenorientierter Schnittstellen. GI-Arbeitsseminar, München 1979. Informatik-Fachberichte 18. Berlin-Heidelberg-New York, Springer-Verlag 1979. X, 230 S., 62 Abb., 5 Tab., DM 25,–, US $ 13.80. ISBN 3-540-09618-3

Der Seminarbericht enthält die vollständigen Vorträge des Arbeitsseminars. Im Aufruf zu den Vortragsanmeldungen wurde die Zielstellung des Seminars in folgender Weise formuliert: „Durch eine Virtualisierung der Apparatur, d. h. durch Nachbildung und Vervielfachung der maschinenorientierten Schnittstellen, können mehrere, evtl. auch unterschiedliche Betriebssysteme parallel in einer Anlage ablaufen. In den vergangenen Jahren war ein zunehmender Einsatz solcher virtuellen Maschinen zu beobachten. Gründe hierfür sind u. a.:
Bei Anwendern wird die Umstellung auf neue Betriebssysteme und/oder neue Anlagen teilweise erheblich erleichtert.
Bei Herstellern kann der Test neuer Betriebssysteme effizienter durchgeführt werden.
Weitere Anwendungen sind absehbar, so daß in vielen Fällen neu zu entwickelnde Rechnerstrukturen von den Anforderungen der virtuellen Maschinen, insbesondere hinsichtlich deren Effizienz, beeinflußt werden.
Die Ziele dieses Arbeitsseminars sind deshalb die Klärung der mit dieser Betriebsweise zusammenhängenden theoretischen und praktischen Probleme, der Austausch von Erfahrungen bei konkreten Anwendungen, sowie eine Diskussion der Entwicklungstendenzen ...". Gegenwärtig liegt der Schwerpunkt der Beschäftigung mit virtuellen Maschinen bei den Herstellern und bei den Nutzern von EDVA. Wesentliche theoretische Untersuchungen fehlen noch. Gegenstand der Vorträge waren hauptsächlich das VM 370 und IBM und das SIM-BS1000 der Siemens AG.

Im Ergebnis des Arbeitsseminars kann eingeschätzt werden, daß sich die Einsatzgebiete, begünstigt durch die gegenwärtigen technologischen Entwicklungen, insbesondere durch die Möglichkeiten der Mikroprogrammierung, laufend erweitern werden und daß die virtuellen Maschinen sogar Einfluß auf die Architektur neuer Rechnergenerationen nehmen werden. Gegenwärtige wichtige Anwendungen sind die Erweiterung von Systemen, ohne die vorhandenen Programme ändern zu müssen, die Testung neuer Betriebssysteme, die Unterstützung der Einführung neuer Systemversionen, die Schaffung unabhängiger und extrem isolierter Subsysteme zur Berücksichtigung des Datenschutzes und der Ausfallsicherheit.

Eine zukünftig immer wichtiger werdende Entwicklungsrichtung sind Netze virtueller Maschinen auf Netzen realer Rechner.

Der Seminarbericht gibt einen repräsentativen Einblick in den gegenwärtigen Entwicklungsstand und macht Motive und Richtungen zukünftiger Entwicklungen deutlich.

Dresden H. Reichel

Fetzer, A. / Fränkel, H., Mathematik. Lehrbuch für Fachhochschulen, Band 3. Hannover-Dortmund-Darmstadt-Berlin, Hermann Schroedel Verlag KG 1979. 423 S., 260 Abb., 11 Tab., DM 44,80. ISBN 3-507-99047-4

Die ersten beiden Bände dieses Lehrbuches werden in der ZAMM 59 (1979), S. 588 besprochen. Der vorliegende dritte Band schließt das Lehrbuch, welches hauptsächlich für Studenten und Dozenten der technischen Fachrichtungen an Fachhochschulen geschrieben ist, ab. In dem ersten Kapitel „Reihen" wird für Zahlen-, Potenz- und Fourierreihen neben den notwendigen Definitionen und Sätzen eine Vielzahl von Beispielen behandelt. Das zweite Kapitel „Funktionen mehrerer Variablen" geht vor allem auf die Differentiation von Funktionen mehrerer Variablen, auf Bereichs- und auf Linienintegrale ein, dabei ist der Tatsache Rechnung getragen, daß erfahrungsgemäß die geometrische Interpretation mehr Schwierigkeiten bereitet als die Technik des partiellen Differenzierens. Auf Anwendungen wie Fragen der Ausgleichsrechnung, Berechnung von Schwerpunkten und Trägheitsmomenten, Coulombfelder, Schraubenlinien und auf Divergenz und Rotation in Strömungs- und Elektrizitätslehre wird ausführlich eingegangen. In einem kurzen dritten Kapitel „Komplexwertige Funktionen" wird die Anwendung in der Wechselstromtechnik behandelt. Das letzte Kapitel „Gewöhnliche Differentialgleichungen" bringt Lösungsmethoden für eine Reihe von typischen Differentialgleichungen erster Ordnung, für lineare Differentialgleichungen zweiter Ordnung mit konstanten Koeffizienten (z.B. die Methode der Laplacetransformation) und für Systeme linearer Differentialgleichungen erster Ordnung mit konstanten Koeffizienten, an vielen Beispielen aus Mechanik, Physik und Elektrotechnik erläutert.

Wie bei den beiden ersten Bänden ist die drucktechnische Gestaltung ausgezeichnet. Das Buch kann durch seine glückliche Kombination theoretischer Grundlagen mit einer Fülle von Beispielen und Aufgaben als Einführung in die höhere Mathematik empfohlen werden.

Berlin G. Schmidt

Jain, M. K., Numerical Solution of Differential Equations. New Delhi-Bangalore-Bombay-Calcutta, Wiley Eastern Limited 1979. XIII, 443 S., 45 Abb., 49 Tab., £ 7.50, US $ 17.65. ISBN 0852264275

This book contains quite a number of good pages, e.g., where formulae are derived in a readable, clear-cut manner. But there are severe lacks, too. We mention some of them occuring in the chapters on partial differential equations:
There is no use of a general stability theory based on the energy method and apriori estimates (see Samarskij). Correspondingly, practicable stability criteria for problems with variable coefficients are not available here. Further, the author, stressing on ADI methods, does not mention such methods as those given by Frjazinov (being second order accurate and unconditionally stable for variable coefficients, in arbitrary dimensions). Nowhere appears the idea that it is important (not only for some nice theory but also for computational efficiency) to conserve properties of the differential equation (selfadjointness, maximum principle ...) in the difference scheme. For elliptic equations, methods like odd-even reduction, fast Fourier transformation, or the Fedorenko-Bakhvalov iteration seem to be unknown. In connection with the Neumann problem, the solvability condition and the necessity of additional data to prevent nonuniqueness are mentioned for the continous formulation, but the discretized equation with singular matrix is said to be solvable.

There is no lack of a foreword (by D. W. Barron) assuring, "that, so far as the numerical solution of differential equation is concerned, that which is not in this book isn't knowledge either."

Berlin G. Stoyan

Gordon, M. J. C., The Denotational Description of Programming Languages. An Introduction. Berlin-Heidelberg-New York, Springer-Verlag 1979. V, 160 S., DM 18,60, US $ 10.30. ISBN 3-540-90433-6

Die denotationale Semantik einer Programmiersprache spezifiziert eine Funktion, die Programme in abstrakte Werte (im allgemeinen Funktionen höherer Ordnung) abbildet. Damit entsteht eine formale Sprachbeschreibung, die Rückschlüsse auf die Korrektheit der Sprachdefinition und Fragen der Implementation erlaubt sowie den Vergleich verschiedener Sprachen anhand eines einheitlichen mathematischen Modells ermöglicht.

Im vorliegenden Buch wird der Leser mit der dabei eingesetzten Beschreibungstechnik (LAMBDA-Notation, Rekursivität) vertraut gemacht. Ausgehend von einfachen Beispielsprachen wird der Leser bis zur denotationalen Beschreibung wesentlicher Konstruktionen solcher Sprachen wie ALGOL 60 und PASCAL geführt. Hauptanliegen des Buches ist es, ein intuitives Verständnis der semantischen Grundlegung von Programmiersprachen zu vermitteln. Die klare und ausführliche Darstellungsweise, die nicht auf mathematische Hintergründe eingeht, wird diesem Ziel ausgezeichnet gerecht. Das Buch kann dem am tieferen Verständnis von Programmiersprachen interessierten Leser, der ohne Vorkenntnisse Einblick in semantische Methoden nehmen möchte, sehr empfohlen werden.

Berlin M. Hartwig

Postlethwaite, I. / MacFarlane, A. G. J., A Complex Variable Approach to the Analysis of Linear Multivariable Feedback Systems. Lecture Notes in Control and Information Sciences 12. Berlin-Heidelberg-New York, Springer-Verlag 1979. IV, 177 S., 53 Abb., DM 21,50, US $ 11.90. ISBN 3-540-09340-0

Die Betrachtung dieser Systeme mit Rückkopplung läuft im wesentlichen auf die Untersuchung von linearen Differentialgleichungssystemen mit konstanten Koeffizienten hinaus: $\dot{x}(t) = \boldsymbol{A} \cdot x(t) + \boldsymbol{B} \cdot u(t)$. Dabei sind $\boldsymbol{A}$ und $\boldsymbol{B}$ Matrizen aus reellen Konstanten, $u(t)$ ein gegebener Vektor („input"), $x(t)$ ein gesuchter. Nach Anwendung der einseitigen Laplace-Transformation entsteht bekanntlich ein gewöhnliches lineares Gleichungssystem. In dem Buch werden die verschiedenen Formen der Stabilität diskutiert, der Zusammenhang mit Eigenwerteigenschaften u.ä. der zugehörigen Matrizen, die Bedeutung gewisser rationaler Funktionen (und deren Riemannscher Flächen), die als Quotient gewisser Determinanten erscheinen, usw. Da die Darstellung mathematisch anspruchsvoll ist, wird mancher Praktiker mit der Lektüre des Buches einige Schwierigkeiten haben, zumal kaum Beispiele ganz vollständig betrachtet werden. Der Ref. empfindet eine gewisse Einseitigkeit in der Literaturauswahl, da Bezüge zur älteren sowie zur nichtangelsächsischen Literatur, sowie zur Literatur z. B. zur Elektrotechnik oder zur Operatorenrechnung kaum vorkommen. Aber dies entspricht wohl etwas dem Charakter eines solchen Forschungsberichtes, und das Buch ist sicher für den Spezialisten von großem Nutzen.

Halle R. Kühnau

Herman, G. T. (Ed.), Image Reconstruction from Projections. Implementation and Applications. Topics in Applied Physics 32. Berlin-Heidelberg-New York, Springer-Verlag 1979. XII, 284 S., 120 Abb., 10 Tab., DM 98,–, US $ 53.90. ISBN 3-540-09417-2

Die Bildrekonstruktion aus Projektionen hat in der letzten Zeit eine große Bedeutung insbesondere durch die Computer-Tomographie, erlangt. Im vorliegenden Buch wird ihre Anwendung in verschiedenen Gebieten gezeigt. Nach einer Einführung durch Herman und Lewitt in die Problematik der Bildrekonstruktion aus Projektionen behandelt Rowland in mathematisch ansprechender Weise die Implementation und die Anwendung der Bildrekonstruktion, die letztlich auf eine Inversionsformel von Radon aus dem Jahre 1917 aufbaut. Hierzu werden effektive Rekonstruktionsalgorithmen für die Verarbeitung großer Datenmengen gebraucht. Es werden Formeln hergeleitet, die die Grundlage für den Convolutions-Algorithmus sind. BRACEWELL zeigt die Anwendung der Bildrekonstruktion in der Radioastronomie. Das von den Radioastronomen für die Bildrekonstruktion benutzte Projektionstheorem wird erklärt. ALTSCHULER berichtet über ihre Anwendung in der Solarphysik. Es werden Methoden beschrieben, mit welchen die globale Verteilung des coronalen magnetischen Feldes und der coronalen Elektronendichte zuverlässig und mit hoher Auflösung erhalten werden können. BUDINGER und Mitarbeiter beschäftigen sich in ihrem Beitrag mit den Methoden der Emissions-Computer-Tomographie, deren Ziel die quantitative Berechnung von physiologischen und biochemischen Prozessen im menschlichen Körper ist. Hierzu werden nach der Injezierung von radioaktiven Chemikalien oder Elementen und deren Zerfall die Strahlung aufgezeichnet und die Bilder rekonstruiert. WOOD und Mitarbeiter befassen sich mit den Problemen der Röntgenstrahl-Transmission-Computer-Tomographie zur Darstellung von veränderlichen Vorgängen, wie z.B. dem schlagenden Herz. Hierbei müssen enorme Mengen von Daten in kurzer Zeit verarbeitet werden. Die möglichen Anwendungen dieses sehr vielversprechenden und bedeutungsvollen Verfahrens werden diskutiert.

Berlin F. GRUND

Marchuk, G. I. (ed.), Modelling and Optimization of Complex System. Proceedings of the IFIP-TC 7 Working Conference Novosibirsk, USSR, 3—9 July, 1978. Lecture Notes in Control and Information Sciences 18. Berlin-Heidelberg-New York, Springer-Verlag 1979. VI, 293 S., 59 Abb., 5 Tab., DM 28,50, US $ 15.70. ISBN 3-540-09612-4

These Proceedings contain most of the papers presented at the IFIP TC-7 Working Conference on Modelling and Optimization of Complex Systems held in Novosibirsk on 3—9 July, 1978. The papers show new results in optimal control and mathematical optimization and their application to modelling, identification, and control of large systems. For instance, in about 10 papers included in the Proceedings mathematical models in immunology (immune response models) are considered. The second large part of papers, (15), is connected with the optimization of complex systems and some applications such as socio-economic-systems or the optimal design of a magnet. In many of these papers computational methods are included.

Leipzig H.-J. SEBASTIAN

Cea, J., Lectures on Optimization — Theory and Algorithms. Notes by M. K. V. Murthy. Tata Institute of Fundamental Research Lectures on Mathematics and Physics 53. Berlin-Heidelberg-New York, Springer-Verlag 1978. V, 236 S., DM 18,—, US $ 8.—. ISBN 3-540-08850-4

Das Buch ist der nichtlinearen Optimierung in normierten linearen Räumen gewidmet und will den Leser sowohl mit theoretischen Grundkenntnissen über Optimierungsaufgaben als auch mit numerischen Verfahren zur Lösung von Optimierungsaufgaben bekannt machen. Das gleiche Ziel verfolgte der Verfasser schon 1971 mit dem im Pariser Verlag Dunod erschienenen Buch „Optimisation: Théorie et algorithmes". Vergleicht man die Titel der beiden Bücher, so könnte man glauben, daß es sich beim vorliegenden Band nur um eine Übersetzung ins Englische handle. Das trifft nicht zu, jedoch sind Aufbau und Inhalt der beiden Bücher einander ähnlich. Einige Ergebnisse aus dem französischen Buch fehlen im neuen Buch vollständig, andere werden auf neue Art behandelt oder durch neue Resultate ersetzt.

Das Buch ist in sechs Kapitel gegliedert. Nach einer einführenden Darstellung der Differentialrechnung in normierten linearen Räumen im ersten Kapitel folgen im zweiten Kapitel Untersuchungen über die Existenz, Eindeutigkeit und Charakterisierung von Minimalstellen reellwertiger Funktionale, sowohl mit als auch ohne Konvexitätsvoraussetzungen. In den folgenden beiden Kapiteln werden iterative Verfahren zur Lösung von Optimierungsaufgaben behandelt und zwar im dritten Kapitel Verfahren für Aufgaben ohne Nebenbedingungen und im vierten Kapitel für Aufgaben mit Nebenbedingungen. Im nächsten Kapitel wird eine auf der LAGRANGE-Funktion basierende Dualitätstheorie entwickelt. Im letzten Kapiteln werden als Anwendungen der in den vorhergehenden Kapiteln gebrachten Ergebnisse optimale Kontrollprobleme und optimale Designprobleme behandelt.

Das Buch, zu dessem Studium Kenntnisse der Funktionalanalysis in normierten linearen Räumen benötigt werden, wendet sich vor allem an Leser, die sich mit der Theorie der Optimierungsaufgaben befassen, und weniger an diejenigen, die sie in der Praxis anwenden, denn es überwiegen die theoretischen Betrachtungen. Selbst die in den Text eingebauten zahlreichen Beispiele sind im Prinzip theoretischer Natur. Numerische Beispiele und Bemerkungen über die praktische Wirksamkeit der einzelnen Verfahren fehlen leider vollständig.

Cluj-Napoca W. BRECKNER

Gihman, I. I. / Skorohod, A. V., Controlled Stochastic Processes. Translated from the Russian. Berlin-Heidelberg-New York, Springer-Verlag 1979. VII, 237 S., DM 64,—. US $ 35.20. ISBN 3-540-90410-7

Im Vorwort zur russischen Ausgabe des 2. Bandes ihres bekannten Werkes „Die Theorie stochastischer Prozesse" bekunden I. I. GIHMAN und A. V. SKOROHOD die Absicht, steuerbare stochastische Prozesse im 3. Band zu behandeln. Doch erscheint dieser 1975 ohne die angekündigte Steuerungsproblematik. Es waren wohl nicht nur Platzgründe, welche die Autoren bewogen haben, „Die steuerbaren stochastischen Prozesse" als separate Monographie 1977 im Kiewer Verlag „Naukova Dumka" herauszubringen (nicht im Moskauer Verlag „Nauka" — wie im Impressum der vorliegenden englischen Fassung fälschlich vermerkt ist), obgleich Anliegen und Diktion mit der des Dreibänders dahingehend übereinstimmen, mathematische Untersuchungsmethoden und Aussagen über stochastische Prozesse bzw. ihre Steuerung lehrbuchartig mit vollständigen Beweisen zu vermitteln, auf konkrete Anwendungen aber von vornherein zu verzichten. Die Abtrennung ist vor allem durch die Spezifik der entscheidungstheoretischen Fragestellung und durch den unterschiedlichen Entwicklungsstand der Theorie steuerbarer stochastischer Prozesse begründet, deren Anfänge erst etwa zwanzig Jahre zurückliegen. Innerhalb dieser jungen mathematischen Disziplin bildet die diskrete stochastische dynamische Optimierung einen relativ weit ausgebauten Zweig (vgl. die Bücher von E. B. DYNKIN, A. A. YUSHKEVICH (1975/79) und D. P. BERTSEKAS, S. SHREVE (1978)), desgleichen die über stochastische Differentialgleichungen beschriebenen steuerbaren Diffusionsprozesse (vgl. W. H. FLEMING, R. W. RISHEL (1975) und N. V. KRYLOV (1977)). Eine einheitliche Theorie steuerbarer stochastischer Prozesse fehlte bisher. Die Autoren versuchen mit der vorliegenden Monographie diese Lücke zu schließen. Auf der Grundlage des Begriffs eines steuerbaren stochastischen Objekts gelingt es ihnen, sowohl eine allgemeine Theorie steuerbarer Prozesse in diskreter Zeit (im ersten Kapitel) als auch in stetiger Zeit (2. Kapitel) sowie gesteuerter stochastischer Differentialgleichungen (3. Kapitel) zu entwickeln. Die benutzte Methode, die auf einer Arbeit von A. V. SKOROHOD (1972) beruht, ist für den stetigen Fall von großer Bedeutung, da sie mit relativ geringem Aufwand zu entscheidungstheoretischen Resultaten führt.

Wenn auch die eingesetzten maßtheoretischen und funktionalanalytischen Hilfsmittel nicht allzu umfangreich sind, ist die Lektüre der interessanten GIHMAN-SKOROHODschen Monographie keineswegs leicht, für Spezialisten der Steuerung dynamischer Systeme bei Unsicherheit aber dringend zu empfehlen.

Leipzig H.-J. GIRLICH

In der DDR beziehbar im Rahmen des Kontingents.

ROLF UNBEHAUEN

Systemtheorie

Eine Darstellung für Ingenieure

3., völlig überarbeitete und wesentlich erweiterte Auflage
1980. 381 Seiten — 149 Abbildungen — gr. 8° —
Lederin 58,— M
Bestell-Nr. 762 655 3 (5722)

Die Systemtheorie hat sich in den vergangenen Jahren zu einem Grundlagengebiet für verschiedene technische Disziplinen entwickelt, insbesondere für die Nachrichtentechnik und die Regelungstechnik. Während in früheren Jahren die betreffenden Disziplinen oft unabhängig voneinander ihr eigenes theoretisches Werkzeug entwickelten, ist es der Systemtheorie gelungen, einheitliche Begriffe und Methoden zu schaffen, die infolge ihrer Loslösung von einer unmittelbaren Bezogenheit auf ein bestimmtes Fachgebiet das Wesentliche erkennen lassen und die verschiedenartigsten technischen Anwendungen ermöglichen. Das Buch trägt dieser Entwicklung innerhalb der technischen Wissenschaften insofern Rechnung, als der Versuch gemacht wird, die wichtigsten Ergebnisse der Systemtheorie im Bereich der determinierten kontinuierlichen Prozesse darzustellen.

Bestellungen durch eine Buchhandlung erbeten

AKADEMIE-VERLAG

DDR-1080 Berlin, Leipziger Str. 3—4

www.ingramcontent.com/pod-product-compliance
Lightning Source LLC
LaVergne TN
LVHW081350011026
849161LV00013BB/3037

9783112649831